Oxford Medical Publications

ESSENTIALS OF OPHTHALMIC OPTICS

Essentials of OPHTHALMIC OPTICS

E. W. ALLEN
Principal Optician, Oxford Eye Hospital,
Oxfordshire Area Health Authority (Teaching)

OXFORD
OXFORD UNIVERSITY PRESS
NEW YORK TORONTO

1979

Oxford University Press, Walton Street, Oxford OX2 6DP

OXFORD LONDON GLASGOW
NEW YORK TORONTO MELBOURNE WELLINGTON
KUALA LUMPUR SINGAPORE JAKARTA HONG KONG TOKYO
DELHI BOMBAY CALCUTTA MADRAS KARACHI
IBADAN NAIROBI DAR ES SALAAM CAPE TOWN

British Library Cataloguing in Publication Data

Allen, E W
Essentials of ophthalmic optics. – (Oxford medical publications).
1. Ophthalmology
I. Title
617.7 RE46 79-40273
ISBN 0 19 261173 9

Set by Hope Services, Abingdon
and printed in Great Britain by
J. W. Arrowsmith Ltd., Bristol

FOREWORD

MONTAGUE RUBEN, F.R.C.S., D.O.M.S.
Consultant Opthalmologist, Moorfields Eye Hospital, London

Mr. Edward W. Allen has devoted a long professional career to ophthalmic optics as applied to the hospital eye service. In that time he has been called upon to teach the optical treatment of the defects of vision to several generations of trainee ophthalmologists and nurses. This requires a background of knowledge that covers a wide field not only in the basic and applied sciences but also of the eye and vision. Very often the trainee ophthalmologist and the general medical, paramedical, and nursing professions find the large and advanced textbooks too difficult and time consuming to use, and at the other extreme simple texts on ophthalmic optics are insufficient in detail. Mr. Allen's background of practical experience presented in this textbook with suggestions for further reading will ensure that the reader can use a book that will stimulate as well as satisfy their interest in this subject.

PREFACE

This book is the outcome of a series of lectures to doctors, nurses, and orthoptists given by the author following his appointment as Senior Optician, subsequently Principal Optician, at the Oxford Eye Hospital. The book is not intended to be a comprehensive volume of ophthalmic optics: many excellent textbooks have already been published on the subject which make up an essential part of the reading for any university course. It was thought that a book presenting an outline of the whole subject in a different style would be of value. The essential aspects of ophthalmic optics are each dealt with concisely with references for further reading. Read as a whole the book serves to correlate the technical with the professional aspects of ophthalmic optics. It is divided into three parts: the preliminary chapters, which cover the main optical principles of the subject are followed by chapters on the visual professional principles. Brief chapters on subnormal visual aids and contact lenses are also included.

The book is intended to be of particular value to the medical profession and those professions closely allied to ophthalmology. Members of the medical profession specializing in ophthalmology will find that the optical aspects of the course leading to the Diploma of Ophthalmology are fully covered. All the essential optical requirements for the Diploma of the Board of Orthoptists are also included. Nursing staff studying for the Ophthalmic Nursing Diploma will find this an easily read volume and reference book with helpful suggestions for further reading. The dispensing optician in training and following qualification will also find the book valuable as a reference work. The student of ophthalmic optics will find the book serves as an introduction to his university course and a concise reference work.

Oxford 1978 E.W.A.

CONTENTS

Contents

ACKNOWLEDGEMENTS

I would like to acknowledge the help given to me by my colleagues past and present at the Oxford Eye Hospital for their suggestions in the production of this book; Miss Tickner, S.R.N., of Moorfields Eye Hospital and Mr. P. Awdry, F.R.C.S., of the Oxford Eye Hospital, for reading the original draft; and finally Miss Elisabeth Orde-Browne for her untiring efforts in producing the final typescript, and the staff of the Oxford University Press for their assistance.

PART I: PURE OPTICS

1 THE NATURE OF LIGHT

Before one can consider the properties of light, such as the mechanism of focusing, reflection, and so on, a brief resumé of its nature and theories of transmission is important. The whole subject is extremely complex and can be dealt with fully only within the realms of higher physics; only a simple conception of the nature of light will therefore be attempted.

Probably one of the best-known facts about light is the speed at which it travels – 186 000 miles per second (this figure is accurate enough for most purposes: more accurate values will be found in most textbooks on optics).

Light may be produced in a number of ways, one of the most common of which is by incandescence: a material is heated to such an extent that light is produced. This light is composed more or less of a mixture of all colours of the spectrum depending on the composition of the material used.

Light is known to have energy, and light falling on an object can be shown to move that object; since one cannot have energy without mass it follows that light must have mass. This phenomenon led to the corpuscular theory of the nature of light which was put forward by Newton in 1704. Earlier than this, however, it had been shown by Huygens that certain phenomena of the nature of light could be explained only in terms of light having a wave motion. Huygens's theory–the undulatory or wave theory of light–was put forward in 1690. Until Einstein and Planck formulated their quantum theory in 1901, the theories of Huygens and Newton were predominant. The essence of the quantum theory is that light consists of a series of energy particles or *quanta* that are promulgated in a wave motion. The quantum theory is, in fact, a compound of the corpuscular and undulatory theories of Newton and Huygens. A more complex treatment of the quantum theory, which is now accepted, is beyond the scope of this book.

Light is only one of a large number of radiations we can detect – and all travel at the same speed of 186 000 miles per second. These radiations include the cosmic, electronic, X-ray, 'visible' light, radio, and electromagnetic waves. The properties of these radiations differ quite considerably, and the radiations themselves require different methods of production.

As these radiations all travel at the same speed their differences must lie in their wavelengths or frequency. (The speed of any wave motion is the product of the frequency and the wavelength.) Within the extremely wide band of radiation, a comparatively small number of radiations are associated with visible light, and we perceive these as the various colours. These range from red to violet, each colour being associated with a particular wavelength or frequency. The wavelengths of these radiations are extremely small and are usually expressed in terms of nanometres (nm):

1 nanometre = 1/1 000 000 000 metre = 10^{-9} metre.

Wavelengths (in nm) for the visible spectrum are

Red	Orange	Yellow	Green	Blue	Violet
760	650	600	550	450	380

These values are approximate and may vary a little from those given in a textbook on optics.

The colour spectrum has been referred to as visible light, and although one of course always thinks of light as something which is visible, light in itself is not visible, as we can appreciate its effect only when it reaches our eyes, or by photoelectric measuring instruments. The so called 'beam' of light which one sees in the cinema, or any heavy atmosphere, is merely light being reflected by the dust particles suspended in the space.

It will be convenient at this point to define the three terms which may be used in connection with the substances through which light may travel.

(1) *Reflection*–A substance is reflective if it sends back all or part of the light incident upon it.

(2) *Absorbtion*–Particular substances can absorb or take up part or all of the light incident upon them.

(3) *Transmission*–This occurs when all or part of the light incident upon a substance passes through it.

No substance has been found as yet which is 100 per cent efficient in any of these properties. We can give an example for each of these properties in turn.

The nature of light

First *reflection*: one of the most highly reflective surfaces is polished silver which, however, reflects only 98 per cent of the light incident upon it, the other 2 per cent being absorbed. *Absorption* is exactly the opposite to reflection. For example, one of the most highly absorbent substances is lamp black–it absorbs approximately 95 per cent of the light incident on it. Absorption or lack of absorption may be easily demonstrated: no matter how black the substance appears to be, it is always possible to form a shadow on it by holding an opaque object between it and the light source. Most so-called transparent substances–glass, water, etc.–*transmit* most of the light incident upon them, but one must not forget the loss which occurs at the substance–air interface–the loss due to reflection.

All three of these properties vary to some extent according to the wavelength of the incident light and the substance. For example an object we see as red is absorbing most of the visible spectrum with the exception of the red wavelengths, which in this case are reflected; red glass is transmitting most of the red light, but absorbs all the other colours.

Modern methods of coating glass can considerably alter the reflective properties of lenses, increasing transmission. One such principle is well known as 'blooming' being used in camera lenses, binoculars, etc.

Rectilinear propagation

When light is travelling in free space it will always travel in a straight line, although upon striking the edge of an opaque medium a certain amount of 'bending' or 'spilling' will take place. This 'bending' in fact means that the shape of an object in a parallel beam of light is slightly diffuse, but has little significance in ophthalmic optics. The property is known as diffraction, which should be referred to in a book on general optics; for this book diffraction will be ignored.

Further Reading

Fincham, W. H. A. and Freeman, M. H. (1974). *Optics* (8th edn). Butterworths, London.

2 REFRACTIVE INDEX

When light enters a different medium from the one in which it is travelling, its speed is altered according to the relationship between the optical density of the two media bounding one another. The optical density of air, or more exactly, vacuum, is defined as unity. The change in speed of light between air and any other substance is referred to as the refractive index of that substance. For example, the speed of light in water is $1\frac{1}{3}$ times slower than that in air; the refractive index of water is therefore 1.33. Various types of glass are made, all having different refractive indices for different uses, the commonest being that of spectacle glass which has a refractive index of 1.523. It will be seen that the refractive index of a substance determines the amount of refraction or deviation of light when it enters a substance of greater or lesser density.

Further Reading

Fincham, W. H. A. and Freeman, M. H. (1974). *Optics* (8th edn). Butterworths, London.

3 REFRACTION OF LIGHT

When rays of light travel through two transparent media, at the boundary some of these rays are reflected and some enter the second medium. The rays entering the second medium are bent or refracted by an amount depending upon the refractive indices of the two media. The refracted rays are always turned towards the normal in the denser medium, or conversely away from the normal in the rarer medium. This refraction is due to the slowing down of the refracted ray in the denser medium.

The amount of refraction is dependent upon the angle of incidence and the refractive indices of the two media. Two laws may be quoted for refraction:

(1) The angle of incidence and the angle of refraction and the normal to the surface at the point of incidence all lie in the same plane.

(2) $$\frac{\text{The sine of the angle of incidence}}{\text{The sine of the angle of refraction}} =$$

$$\frac{\text{The refractive index of the substance containing the refracted ray}}{\text{The refractive index of the substance containing the incident ray}},$$

i.e. $$\frac{\sin i}{\sin r} = \frac{n \text{ of medium } r}{n \text{ of medium } i}.$$

This is *Snell's law* (Fig. 3.1).

It will be appreciated that in theory it would be possible for a ray of light making an angle of inci-

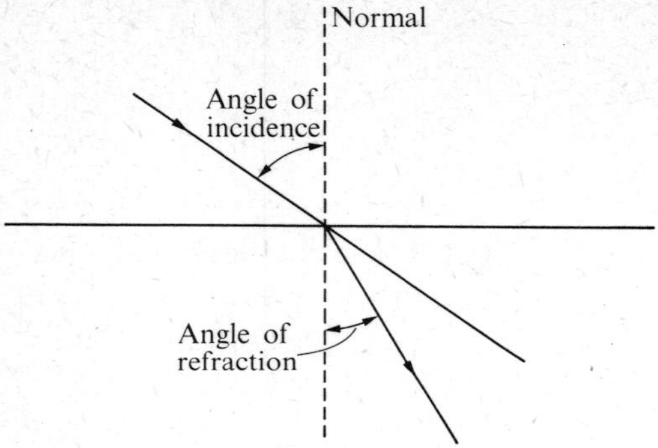

FIG. 3.1. Refraction.

dence at 90° to be refracted, the angle of refraction within the material being the maximum angle possible for that particular material. In practice, of course, this does not happen, as all the light is reflected at the surface.

Assuming now that we have a ray of light passing from a dense to a rare medium, as the angle of incidence is increased (see Fig. 3.2), the angle of refraction in the rare medium also increases: but at the interface a proportion of the rays of light is reflected back according to the laws of reflection. Where the angle of incidence would produce, in theory, an angle of refraction of 90° there is a critical point where any increase in the angle of incidence will produce total internal reflection of the rays of light being reflected at the interface; this angle is known as the *critical angle*. The phenomenon of total internal reflection is used in many optical instruments, particularly for the prisms in binoculars, prismatic loupes, and slit-

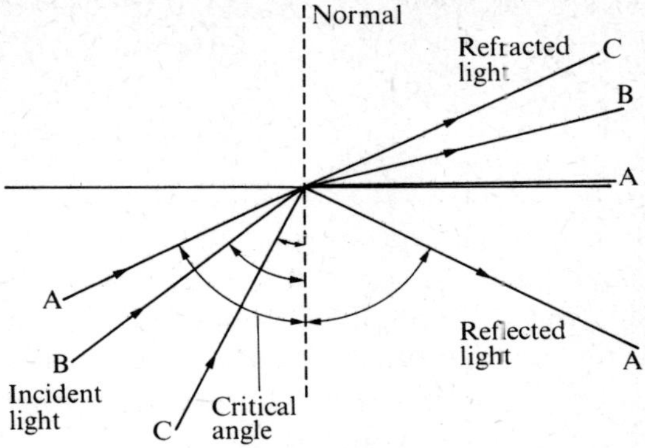

FIG. 3.2. Total internal reflection.

lamp microscopes. A brief account of optical magnification and instruments will be found in Chapter 19.

The angle of refraction is also dependent on the wavelength of the incident beam. White light consists of a mixture of all colours of the spectrum, all the colours having different wavelengths, therefore all refracted by a different amount. This means that a beam of white light is broken up into its spectral colours, or dispersed, immediately after being refracted. This phenomenon may sometimes be noticed with ophthalmic prisms of more than 4 prism dioptres (see Chapter 8 for a discussion of the power of prisms).

Further Reading

Fincham, W. H. A. and Freeman, M. H. (1974). *Optics* (8th edn). Butterworths, London.

4 REFLECTION OF LIGHT

When rays of light are incident upon the surface of a differing medium, some of the light is reflected back into the first medium. This amount of reflection will depend upon the type of surface bonding the two media, which, if opaque and highly polished, may reflect up to approximately 98 per cent of the incident light, as, for example, with polished silver. When the surface is perfectly flat or plane *regular* reflection occurs; with an irregular surface diffuse or *irregular* reflection occurs, as for example, with matt paint, blotting paper, etc. When rays of light are

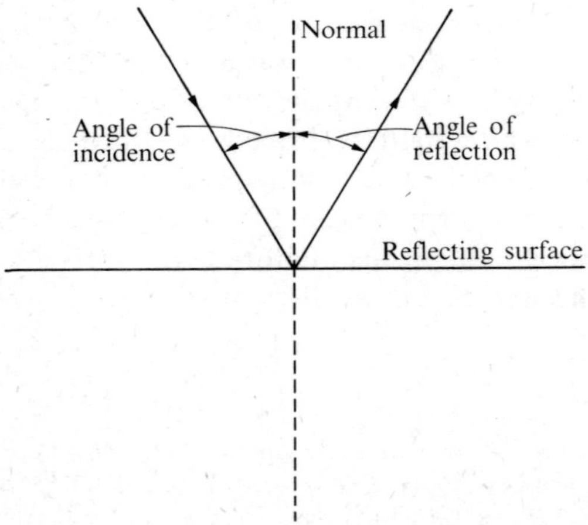

FIG. 4.1. Reflection

regularly reflected, the reflected rays follow the laws of reflection. The laws of reflection state:

(1) The incident, reflected ray, and the normal to the surface at the point of incidence all lie in the same plane.

(2) The angle of incidence and the angle of reflection are equal (see Fig. 4.1)

These laws will apply to any regular surface, whether flat or curved. It will be seen that in the case of a plane mirror that the image of any given object will be formed at exactly the same distance behind the reflecting surface as the object is in front (Fig. 4.2).

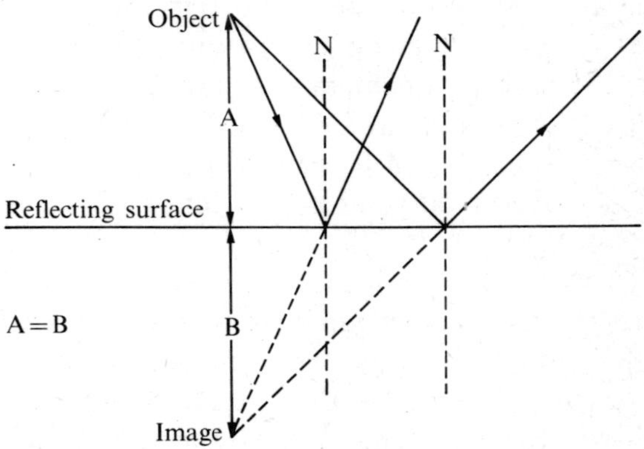

FIG. 4.2.

Further Reading

Fincham, W. H. A. and Freeman, M. H. (1974). *Optics* (8th edn). Butterworths, London.

5 IMAGE FORMATION

Different types of images are encountered in optics and the following is a brief outline of the terms used.

Real images

A real image is formed when rays of light are reflected by a concave mirror or refracted by a convex-lens system (Fig. 7.8). As a result an image is formed in another plane (in front of the mirror or behind the lens) by the rays of light converging to a focus in that plane. This image may be viewed (in the case of the convex lens) by placing an opaque obstacle in the plane (the film in a camera is in this plane) or, alternatively, the image may be viewed by placing the eye beyond the image and focusing on it. Real images are always inverted relative to the object.

Virtual images

A virtual image is formed when rays of light are reflected by a convex or plain mirror or refracted by a concave-lens system. In this case, however, an image is not formed in a plane because the rays of light are diverging away from a plane (on the same side of the lens or behind the mirror) where the image *appears* to be. The only way the image can be viewed is by positioning the eye in the path of the diverging rays. Virtual images are always erect relative to the object. (Virtual images will also be formed when the object is placed at a distance less than the focal length of a concave mirror or convex lens, as will be seen from chapters on mirrors and lenses.)

The distinction between the terms *erect* and *inverted* should also be noted. An erect image is exactly like the object, whereas an inverted image is upside down and laterally reversed. If we consider the letter F; then F is the erect image of F, and Ⅎ is the optically inverted image of F.

Further Reading

Fincham, W. H. A. and Freeman, M. H. (1974). *Optics* (8th edn). Butterworths, London.

6 MIRRORS

The mirrors most commonly associated with optical instruments and ophthalmic optics are of three different types:

(1) Plane – surface is completely flat, that is, the radius of curvature is infinity.

(2) Convex – the surface is curved away from the direction of incident light.

(3) Concave – the surface is curved towards the direction of incident light.

The surface of a convex or concave mirror may be part of a regular sphere, in which they are called spherical mirrors; but where large aperture mirrors are required for producing undistorted images, the surface must be specially worked in the shape of a parabola. Parabolic mirrors are used in reflecting telescopes and the headlamps of cars.

The curvature of a spherical mirror may be expressed in terms of its radius, but often the power of the mirror is expressed in *dioptres*, indicated by the capital letter R, this value being the reciprocal of the radius of curvature in metres. If a concave mirror is set up with parallel rays of light incident upon it the reflected rays will be found to come to an approximate focus at a distance from the mirror equal to about half the radius of curvature. It will be found that the smaller the diameter of the mirror the less distorted this focal point will be, as it is only in the small area surrounding the optical axis of the mirror where these parallel rays of light come to focus at a common point. If a number of parallel rays are drawn in the construction for the reflection of light by a curved mirror, the normal to the surface being the radius at the point of incidence, it can easily be seen how these parallel rays of light move further and further from the focus towards the centre of the mirror, the further the constructed ray moves from the optical axis (Fig. 6.1). This effect is known as spherical aberration.

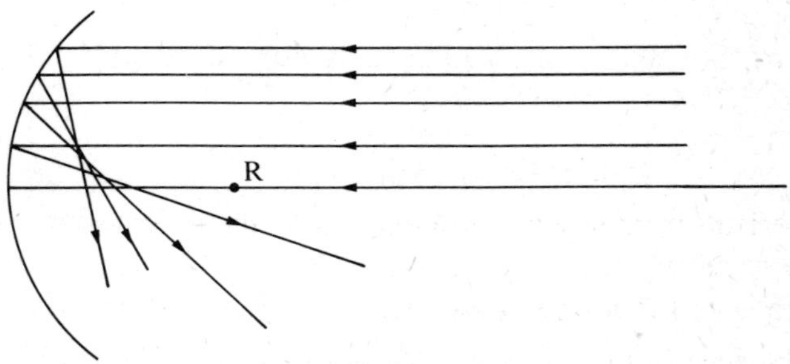

FIG. 6.1. Concave mirror showing spherical aberration.

As mentioned before, the point of focus for rays in an area immediately surrounding the optical axis is equal to almost half the radius of curvature and this point is called the focal point of the mirror. The focal length of the mirror is therefore approximately equal to half the radius of curvature.

A similar diagram may be drawn for a convex mirror. However, in this case the rays of light diverge after reflection from a point behind the reflecting surface, which may be referred to as the focal point of the mirror. The focal length will again be approximately equal to half the radius of curvature (Fig. 6.2). These two diagrams illustrate spherical aberration.

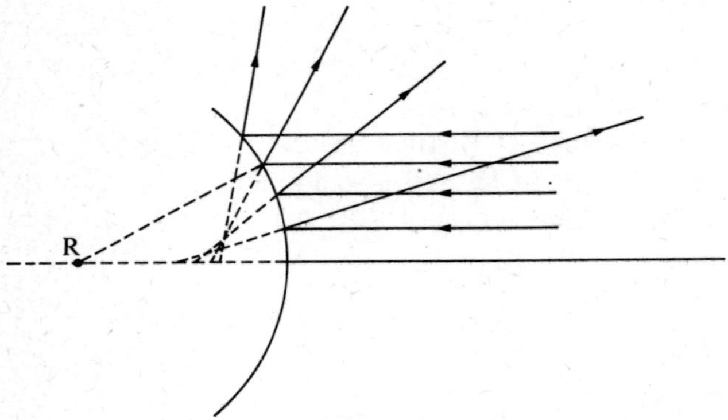

FIG. 6.2. Convex mirror showing spherical aberration.

Formation of images by mirrors*

The image formed by a plane mirror has already been dealt with in conjunction with the laws on the reflection of light, and it was found that a virtual image of equal size and shape at the same distance behind the mirror would be formed.

The formation of images by concave and convex mirrors is more complex, and may be illustrated by a series of diagrams. The construction of any optical image may be undertaken for any type of mirror or

*It will be assumed in the following figures that the mirrors concerned are small so that the spherical aberration can be ignored.

lens, provided at least two different rays are drawn whose course is known after reflection or refraction by the mirror or lens concerned. In the case of mirrors there are three rays whose path may be traced as follows:

(1) Any ray parallel to the optical axis will, after reflection, pass through (or may be assumed to pass through) or appear to be reflected from the direction of the focal point of the mirror.

(2) Any ray of light passing through or towards the centre of curvature will be normal to the surface of the mirror and will therefore be reflected back along its path.

(3) The ray of light passing to or through the focal point will, as seen from (1) be reflected parallel to the optical axis By using any two of these rays the position of an image formed by a mirror may be constructed geometrically. Figs. 6.3–6.5 give a few

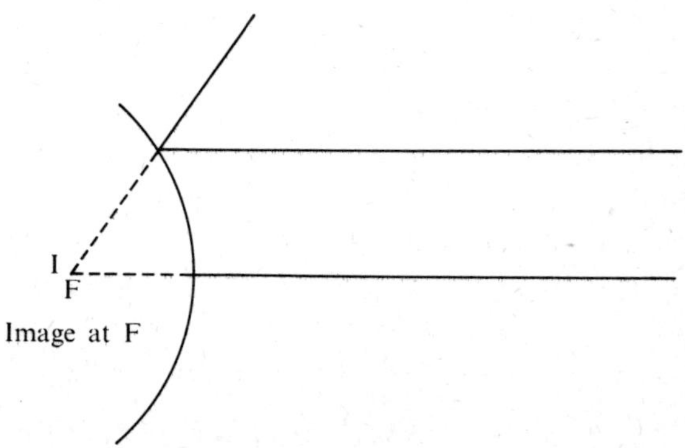

FIG. 6.3. Object at infinity, image at F.

examples of image formation, and also serve to illustrate the position and relative size of the image in comparison with the object.

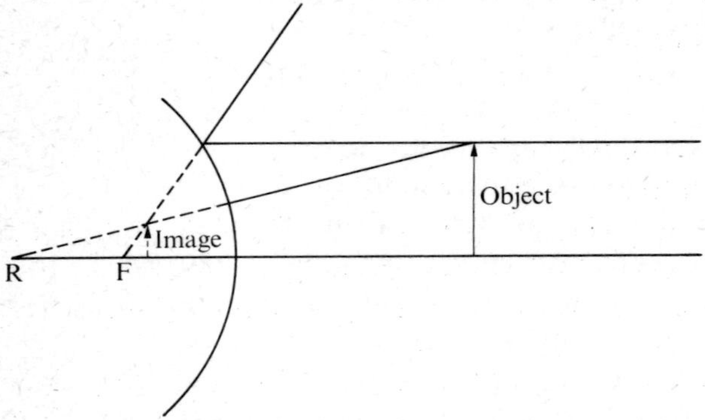

FIG. 6.4. Object finite distance from the mirror.

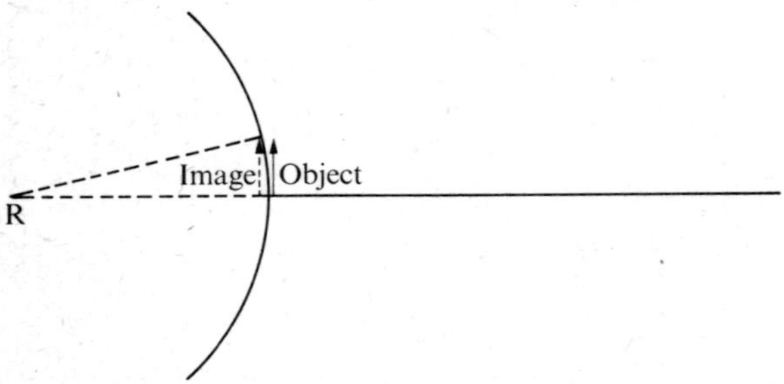

FIG. 6.5. Object at the mirror.

It will be seen that the image in all cases for the convex mirror is erect, virtual, and smaller, except when the object is in contact with the mirror when it will be equal in size. In the case of concave mirrors the image formation is more complex and the image may be much smaller, larger, or virtual. Figs 6.6–6.10 show various image formations using a concave mirror.

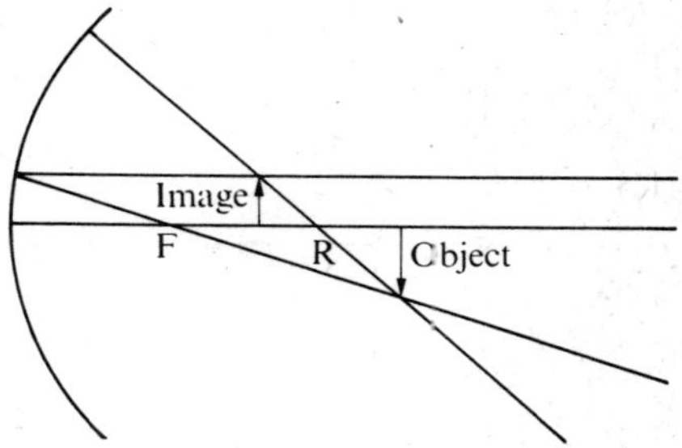

FIG. 6.6. Concave mirrors.

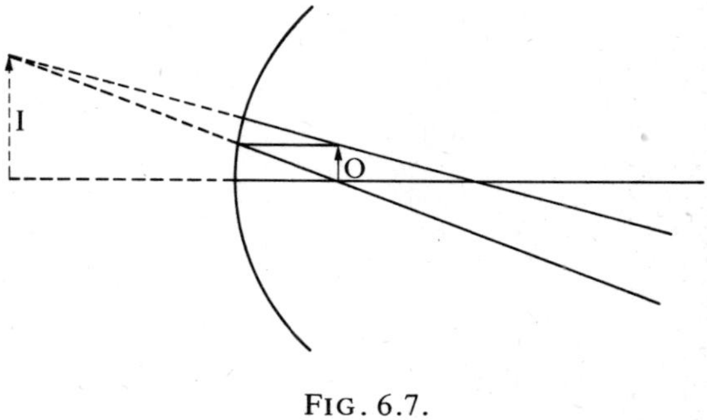

FIG. 6.7.

Many examples come to mind for the everyday use of mirrors, so only a few will be given here. Convex mirrors may often be found incorporated in the wing mirrors of cars and also as reducing reflectors framed for ornamental use. As the view seen in the convex

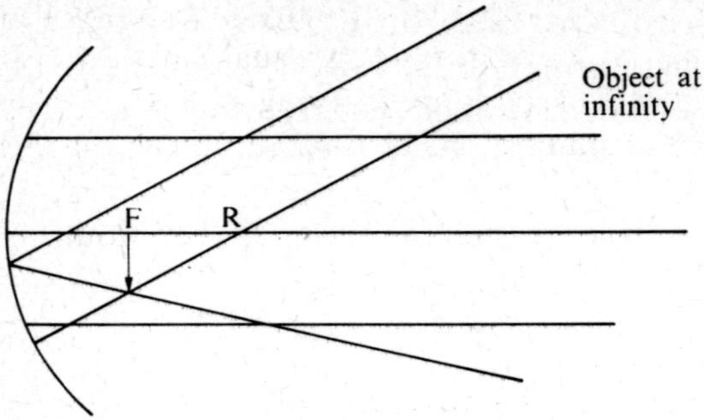

FIG. 6.8.

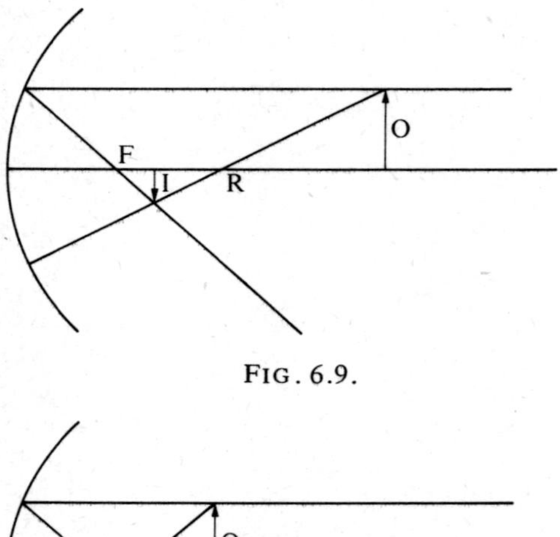

FIG. 6.9.

FIG. 6.10.

mirror is of a wide angle, many mirrors of this type are now seen in stores for visual security. Concave mirrors are used as magnifying mirrors for application of cosmetics, shaving etc., and also in scientific and optical instruments. One of the largest concave mirrors in existence is used in the Mount Wilson observatory, having a diameter of 5.08 m (200 in).

Further Reading

Fincham, W. H. A. and Freeman, M. H. (1974) *Optics* (8th edn). Butterworths, London.

7 LENSES

Lenses may conveniently be grouped into two main classes: (i) those for ophthalmic use, e.g. in spectacles for the correction of ametropia and (ii) those for optical instruments. Both these classes of lenses are for specialist use, but are fundamentally different from each other.

Ophthalmic lenses are rarely compound. They have two surfaces, front and back, and usually one surface is spherical or paraboloid. The second surface for the correction of astigmatism, for example, is cylindrical or toroidal. Other variations in sight defect necessitate special lenses, e.g. bifocal and lenticular lenses.

The lenses for optical instruments are compound in some instances, that is there is more than one lens in the system, and usually all the optical surfaces are part of a spherical curve.

Spherical lenses are those lenses where the power, or focal length, of every meridian (or plane) passing through the optical centre is the same. The back and front surfaces will have a curve of regular spheres, the sizes of which may or may not be equal, depending on the design of the instrument for which they are used. The curvature will be found to vary considerably, even though the focal length may be constant, in lenses which are used for different purposes in different optical instruments. The curvature of lenses in ophthalmic use will be found to follow a fairly regular pattern, which will be dealt with later in this book (Chapter 17).

We must now consider the effect upon rays of light when they pass through a spherical lens. It will be seen from Fig. 7.1 that some rays of light incident

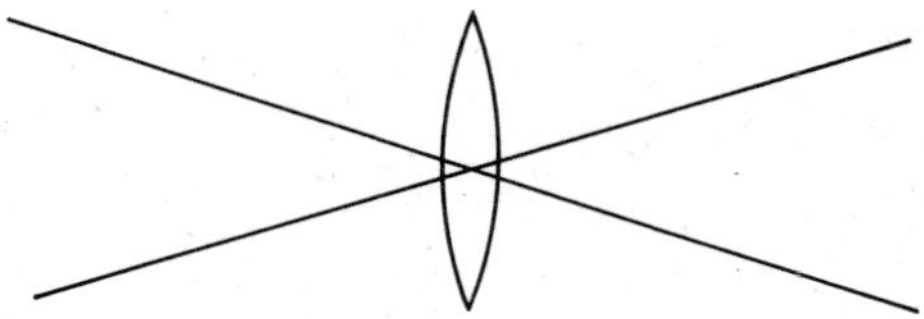

FIG. 7.1.

upon a lens pass through both surfaces without deviation. These rays all pass through the point on the lens known as the optical centre. No deviation takes place at this point, because the lens may be considered as presenting to the ray of light two surfaces parallel to one another and we have already seen, when dealing with refraction, such a ray of light will pass through parallel surfaces undeviated, although in fact it may be slightly displaced. The ray which is normal to both surfaces is the optical axis of the lens. Any ray of light striking the lens elsewhere will be refracted by the lens, the amount of refraction increasing the more the ray of light is displaced from the optical centre. This is because there is an ever-increasing angle between the front and back surfaces, and the light is being refracted as by a prism (see Chapter 8). It will be seen from Fig. 7.2 that parallel rays of light incident upon the lens at points away from the optical axis will all be bent,

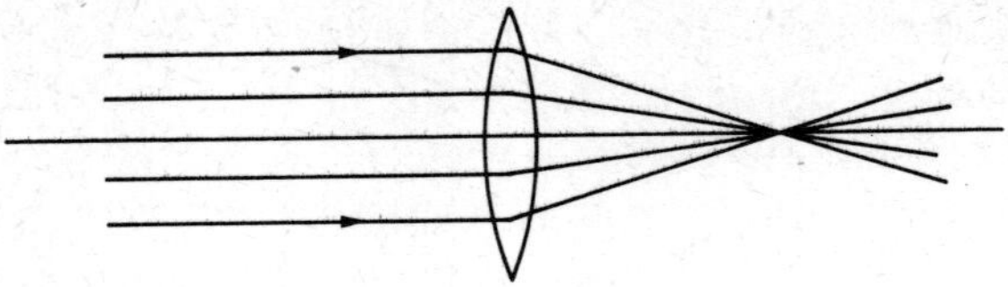

FIG. 7.2.

or refracted, towards a common point or focus. The focus of these rays of light will not be a point in a simple lens owing to the fact that the peripheral rays of light will be refracted to cross the optical axis at a point nearer to the lens than the central rays of light (Fig. 7.3). As was seen with mirrors,

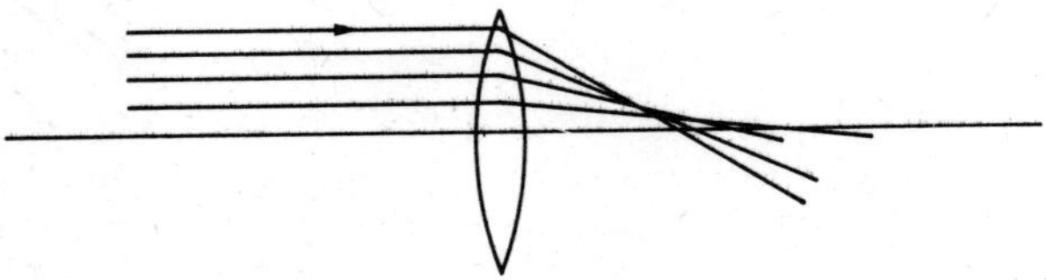

FIG. 7.3. (Showing spherical aberration.)

this spherical aberration usually has to be corrected in lenses of scientific optical instruments. In ophthalmic lenses, spherical aberration is normally of no great significance, and it is in fact corrected to some extent by working the lenses in a curved form.

The diagrams so far have shown the lenses with convex surfaces, the rays of light being converged after refraction. These are known as positive or plus lenses. Lenses with concave surfaces, or minus lenses, are ground with a convex tool and are also of spherical curve, but will diverge the rays of light after refraction, the refracted rays appearing to come from a

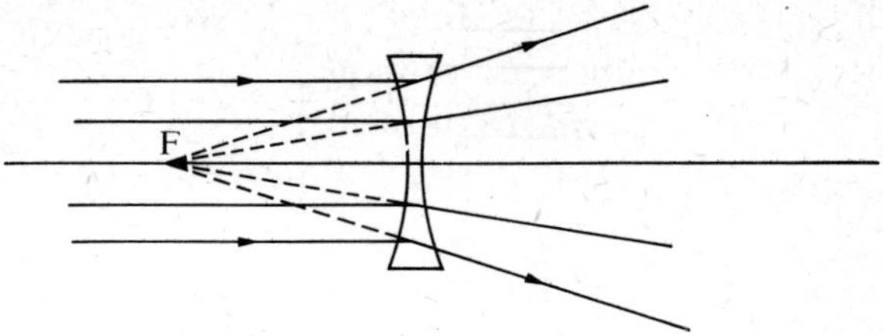

FIG. 7.4.

common point, which is again referred to as the focus, see Fig. 7.4. Spherical aberration will also occur in a simple lens of this sort.

The formation of images by spherical lenses

Positive lenses

From the foregoing explanation of rays of light striking a spherical lens, it will be seen that the position of the image can be constructed by drawing the subsequent course and position of two rays. In actual fact one can draw three rays, if one includes the ray drawn passing through the focal point of the lens before striking it, which will be refracted parallel to the optical axis. Any two of these three rays may then be used to construct the position of an image, given the position of the object and focal points of the lens. Figs 7.5–7.10 show the image formed of an object situated at infinity and then moved progressively nearer to the lens.*

Fig. 7.5 shows the object at infinity, the image

*In all these explanations it has been assumed that a simple lens will produce correctly a point image of a point object, and spherical aberration has been ignored.

being formed at the focal point: the image will be real and inverted, having minimal dimensions. If the object is now brought from infinity towards the lens, the image will increase in size and move further away.

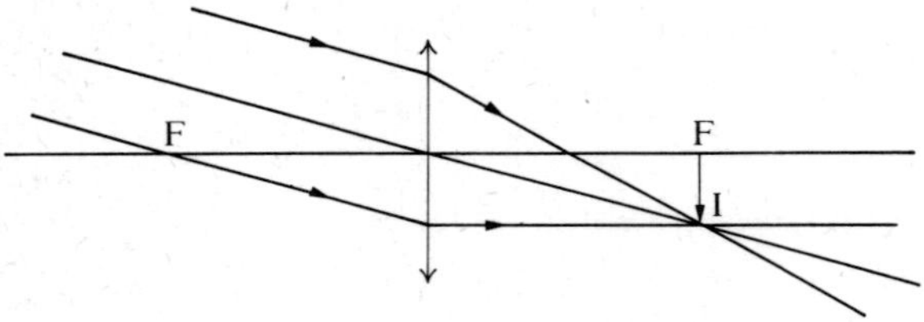

FIG. 7.5.

Fig. 7.6 shows the object situated in a position greater than twice the focal length of the lens and producing a real inverted image, and although bigger than the image at infinity, it will be smaller in size than the object. These two diagrams serve to show the principle of the image formed by a normal photographic camera which has not been adapted for extremely close work.

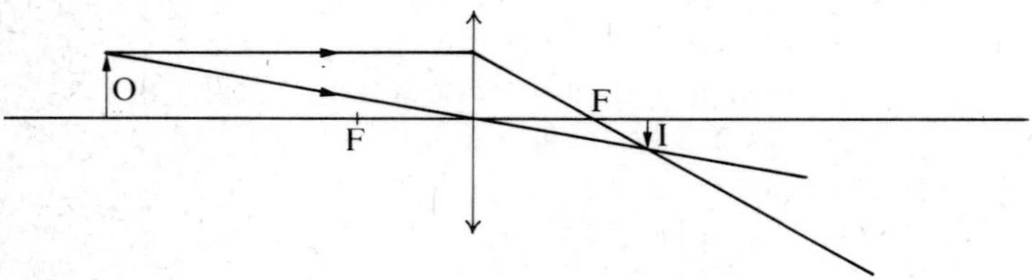

FIG. 7.6.

In Fig. 7.7 the object is now situated at twice the focal length, and the image is also situated at twice the focal length on the opposite side of the lens, and again is real and equal in size, as for instance in photographic copying.

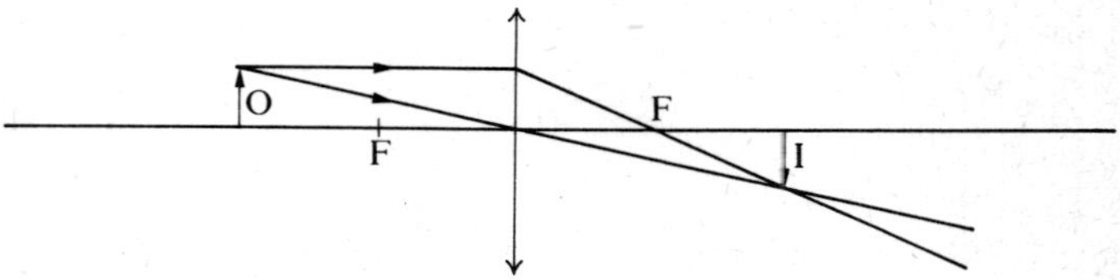

FIG. 7.7.

Fig. 7.8 shows in effect the positions of object and image in Fig. 7.5 reversed, producing a large, real, inverted image. This diagram also illustrates the basic principle of the standard projection lantern.

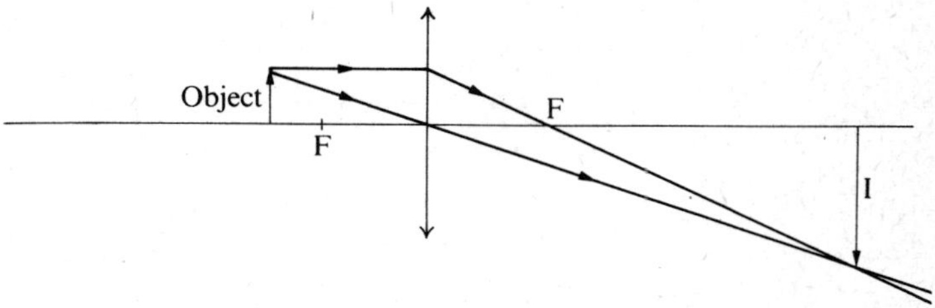

FIG. 7.8.

In Fig. 7.9, with the object situated at the focal point of the lens, rays of light will emerge parallel, and in theory will produce an image at infinity, which will of course have maximum dimensions. This image would be real and inverted. If the eye is placed on the opposite side of the lens from the object in

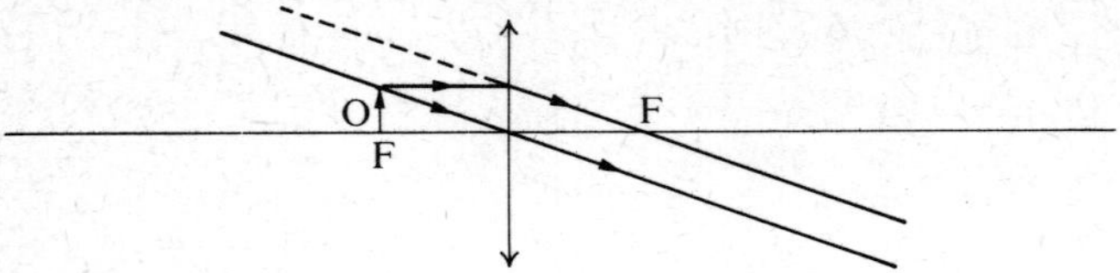

FIG. 7.9.

this instance, parallel rays of light would, in theory, focus on the retina appearing to come from a virtual erect image situated at infinity on the opposite side of the lens. The image would naturally be very much distorted, but if the object is moved towards the lens, as shown in Fig. 7.10, an enlarged image would be

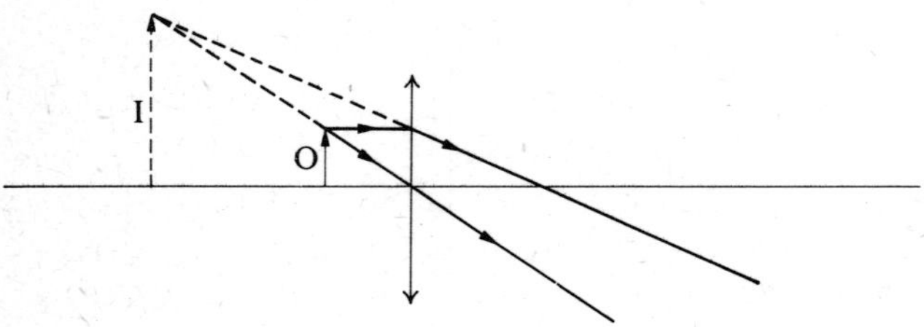

FIG. 7.10.

seen but smaller than the theoretical image of Fig. 7.9. This is the principle of any simple magnifier, and will be referred to again in Chapter 19.

Further movement of the object towards the lens will progressively reduce the image size, till the object is in contact with the lens when, in theory, the object and image would coincide and be of equal size.

Negative lenses

Rays of light will always diverge after passing through a negative lens. The images formed, therefore, will never be real, remaining virtual at all times.

In Fig. 7.11 the object is situated at infinity, the virtual image being formed on the same side of the lens at the focal point, and of minimal dimensions.

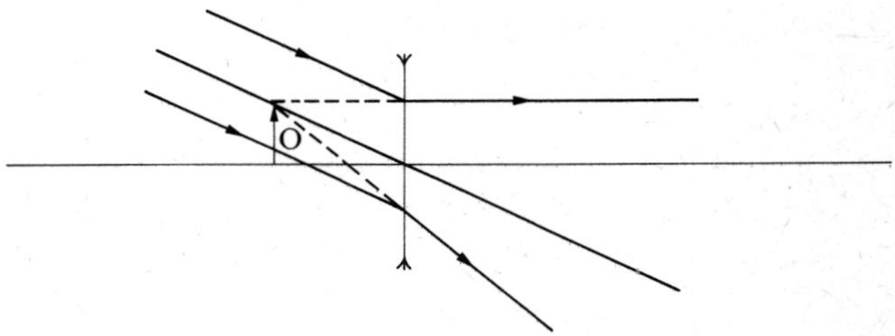

FIG. 7.11.

Figs. 7.12–7.14 serve to show the object being moved progressively nearer to the lens and the image increasing in size, and also being closer to the lens. The object and the image will be of greatest size when the object is in contact with the lens. In these diagrams spherical aberrations have again been ignored.

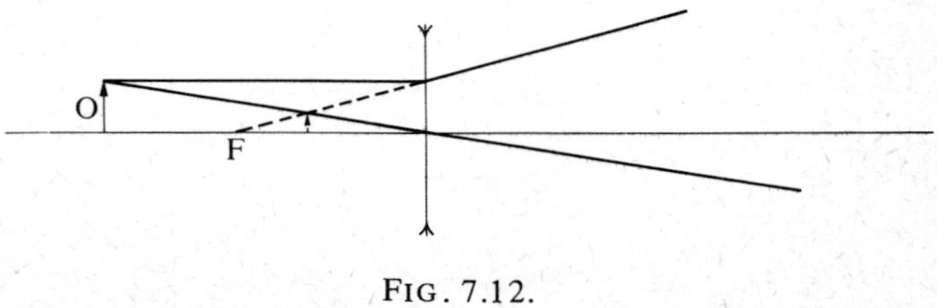

FIG. 7.12.

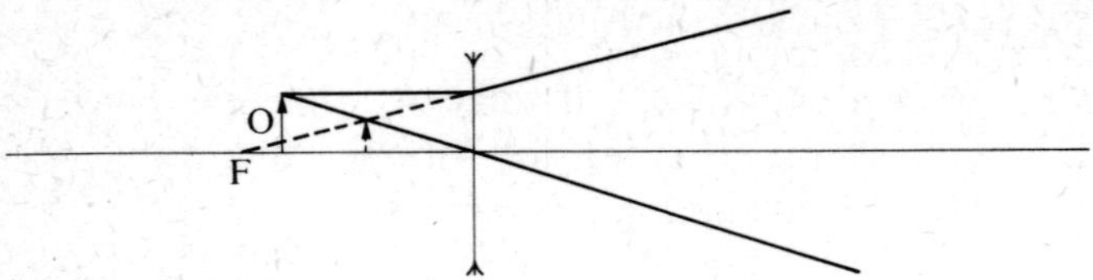

FIG. 7.13.

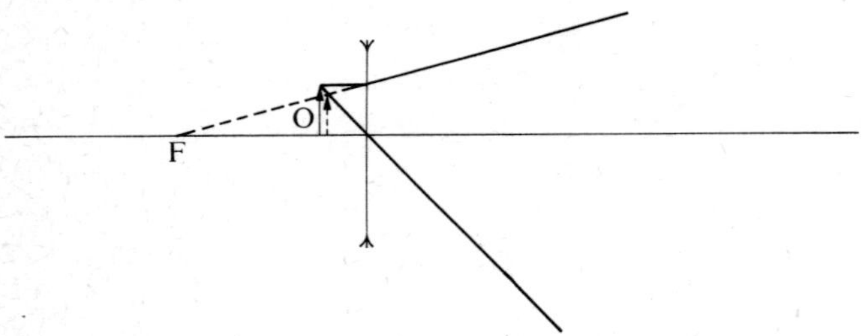

FIG. 7.14.

Depth of focus

When an image has been formed by a large-diameter lens, the rays of light passing through the edge of this lens will be refracted through a greater angle than at the centre, and will cross the optical axis forming a larger angle than the central rays of light. It will be seen from Fig. 7.2 that if the lens is not correctly focused the image will be more blurred the larger the diameter of the lens. Conversely any screen on which the image may be formed, e.g. photographic plates, which allows a greater latitude of movement along the optical axis without any undue blurring is said to have 'depth of focus'. This property is extremely important in photography in deciding the size of the iris diaphragm and focus of the lens. Depth of focus relative to the eye will be dealt with in connection with the measurement of refractive error.

The formation of images by cylindrical lenses

The focus of a cylindrical, or spherocylindrical, lens will vary according to the meridian along which the rays of light pass. It will thus be appreciated that parallel rays of light lying in a plane running through the power meridian will come to a focus at the focal point of the lens in a similar way to that of a spherical lens. If the lens is a planocylindrical lens the rays of light in a plane at right-angles to the power meridian will undergo no refraction at all; and the image, therefore, of any point source will be a line parallel to the meridian of no power. This line is in fact an image of the point source, but blurred or elongated in one meridian only, see Fig. 7.15. When the lens is

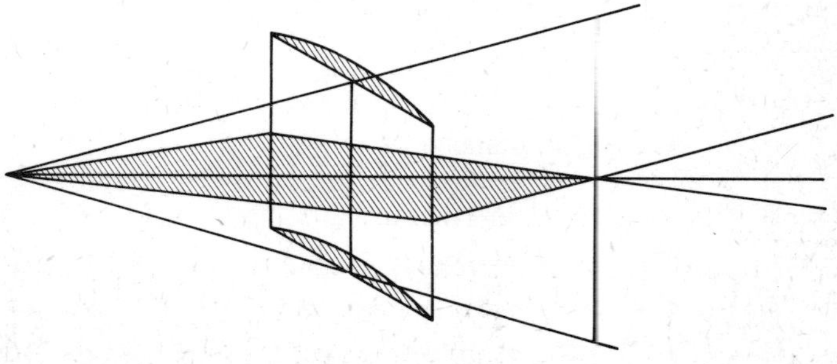

FIG. 7.15.

spherocylindrical, i.e. having power in both meridians, the lens will form two line-images of the point source being at the focal point of each other, see Fig. 7.16.

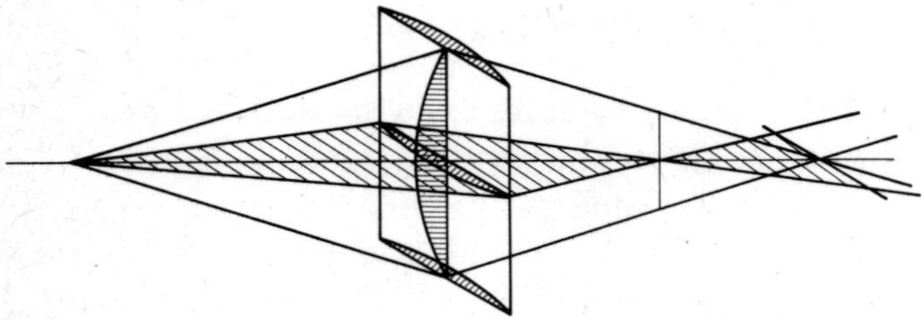

FIG. 7.16.

Power of lenses

The power of lenses may be expressed either in terms of their focal length or, more generally as in ophthalmic use, in dioptres. The power of a lens in dioptres is given by

$$\frac{1}{\text{Focal length in metres}},$$

the appropriate sign + or − indicating whether it is positive or negative lens.

The effect of lenses in contact

When a series of thin lenses are held in contact with one another, their total power is equal to the algebraic sum of their individual powers. Should any of the individual lenses by cylindrical, the resultant power can also simply be found in the same way, providing all the cylindrical lenses have their axes parallel or at right-angles to each other.

When a system of lenses are not held in contact, being separated from one another by a different medium, or media, the sum of their powers will not equal the effective power or focal length of the system. The focal length, or power, of this system can be calculated by tracing the known rays of light for each surface and so establishing the final focal-point. This process is somewhat laborious, and clearly a number of calculations have to be made.

Six cardinal points may be found for any homocentric optical system. These six points may then be used to represent the entire optical system for calculating purposes. The six cardinal points are:

> two focal points corresponding to the focal points of a simple lens, two nodal points loosely corresponding to the optical centre of a simple lens, and two principle points loosely corresponding to the position of a simple lens.

Where all surfaces of a compound system are bound by media having a common refractive index, the nodal points and principle points will coincide. Where the surfaces of the system are bound by media of differing refractive index, the nodal points and principle points will not coincide, but the distance between the principle points will be equal to the distance between the two nodal points. The following diagrams of two simple lenses forming a compound system show the relative position of the six cardinal points. It will be seen that in Fig. 7.17

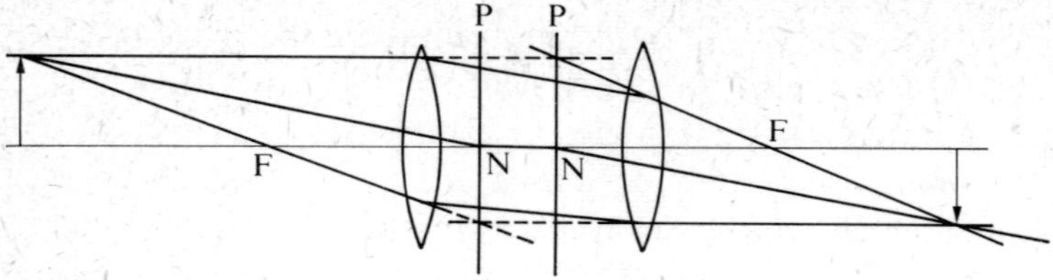

FIG. 7.17.

the nodal points and the principle points coincide, representing the system where all surfaces are bound by the same medium.

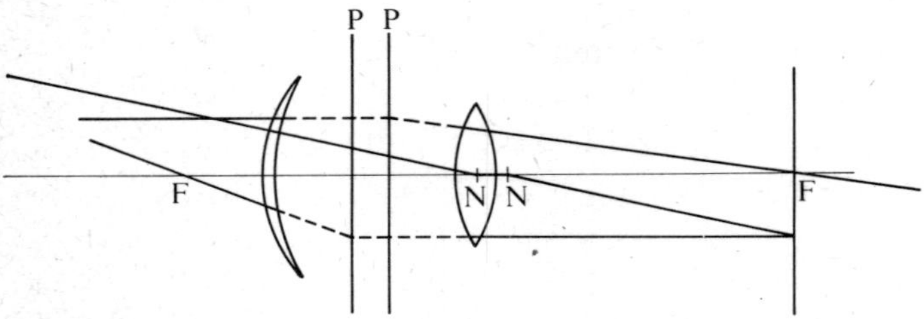

FIG. 7.18.

Fig. 7.18 shows a compound system where one surface is not bound by the same medium as the other surfaces of the system, and in which the cardinal and principle points do not coincide. This last diagram is similar to the diagram which will be dealt with later showing the optical system of the human eye.

The original calculation of the positions of cardinal points was the work of the mathematicians Gauss and Listing.

Lenses for ophthalmic use will be dealt with in the chapter on spectacles.

Further Reading

Fincham, W. H. A. and Freeman, M. H. (1974). *Optics* (8th edn). Butterworths, London.

8 PRISMS

Prisms may be considered as a wedge-shaped piece of transparent material having two faces inclined at an angle towards each other. The point at which these two faces meet is referred to as the apex; the edge furthest from the apex being referred to as the base. A prism may be made having a larger number of faces in which case refraction or reflection may take place at any of these faces. This type of prism is used in specialized optical instruments. The most common form of prism is probably the right-angle prism having three faces, two inclined at 90° to each other and both at 45° to the third face. This type of prism is used as a reflecting prism in the manufacture of prism binoculars.

For ophthalmic use the angle between the two faces is comparatively small and the prism is used for refracting light, probably in conjunction with lenses correcting the patient's ametropia. The amount of refraction which takes place in a prism will depend upon:

(1) the apical angle,
(2) the refractive index, and
(3) the angle of incidence.

Points (1) and (2) are under the control of the manufacturer, being made to a given prescription, but (3) is under the control of the user.

Some reference must be made as to the effect of the prism on oblique rays of light. If a ray of light passes through a prism obliquely, the resultant apical angle will be increased (see Fig. 8.1). The prism should, therefore, always be used, as nearly as possible, so that the rays of light enter it normally, i.e. in a direction at right-angles to the prism surface. In ophthalmic uses in spectacles this is impossible and some prismatic distortion is unavoidable.

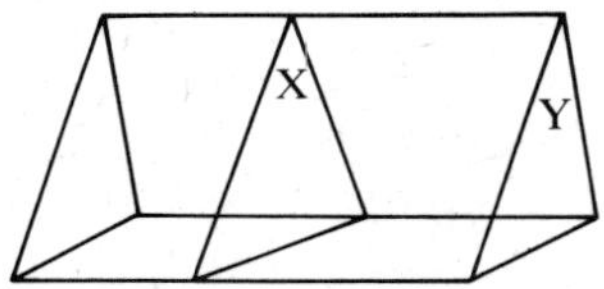

FIG. 8.1. Prism showing distortion due to increased angle X deviating oblique rays through a greater angle.

As regards point (3), if a distant object is viewed and a prism is held in front of the eye and is then rotated about an axis in the plane of the prism, but at right angles to the base apex line, it will be found that the deviation will increase and decrease according to the angle of incidence but will never be less than a certain amount. The angle at which this minimum amount of deviation occurs is referred to as the angle of minimum deviation, and occurs when the rays of light pass through the prism symetrically, i.e. where the angle of incidence and the angle of emergence are equal. The angle of minimum deviation may be found from the formula

$$d = (n - 1)a,$$

where d is the minimum deviation, n the refractive index, and a the apical angle. This formula refers only to prisms of 'small' angles, i.e. those normally for ophthalmic use (Fig. 8.2)

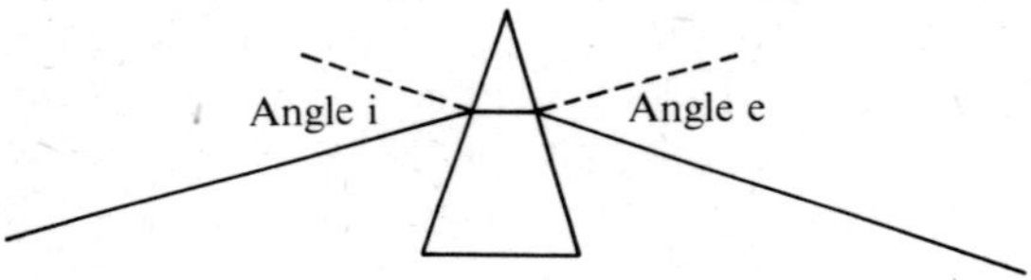

FIG. 8.2. Minimum deviation angle i = angle e.

There are four common systems for denoting the power of a prism:

(1) apical angle,
(2) angle of minimum deviation,
(3) prism-dioptre, and
(4) centrads.

In ophthalmic work the prism-dioptre system is now almost universally used.

(1) *Apical angle.* This system is not now used for ophthalmic prisms. In ophthalmic use it is the deviation produced by the prism in which we are interested and, as we have already seen, is dependent upon the refractive index of the material of which the prism is made. In optical instruments, however, prisms are expressly designed for a particular purpose and are normally classified in terms of the angle of the working faces, and of course the refractive index.

(2) *Angle of minimum deviation.* By measuring the angle of minimum deviation, expressed in degrees,

the apical angle and the refractive index will be taken into account and need only be known to the manufacturer.

(3) *Prism-dioptre system.* A prism having a power of 1 prism-dioptre (1^{Δ}) will deviate a fixed point 1 unit at 100 units along a tangent scale normal to the line of sight of the observer. It will thus be seen that 1 dioptre-prism gives a ratio of 1/100 deviation to distance, and a 3 dioptre prism 3/100. The actual angle of deviation is thus being measured in terms of its tangent. It will be seen, therefore, that the actual amount of deviation in degrees is not linearly related to the deviation in dioptres, but becomes progressively smaller for each increase of prism dioptre (Fig. 8.3).

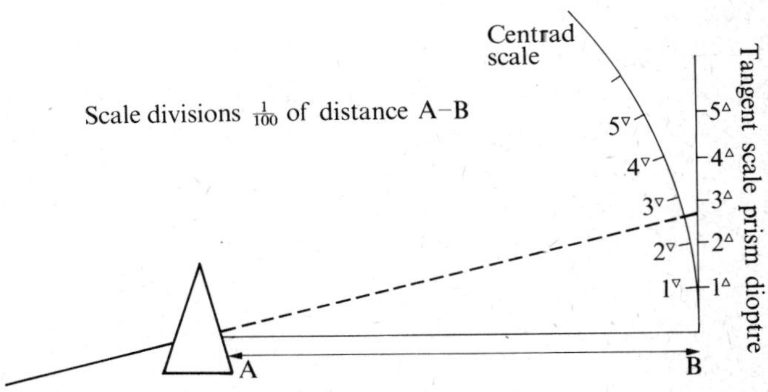

FIG. 8.3. Prism dioptre ($\triangle$) and centrad (∇) scales.

(4) *Centrads.* In order to overcome the non-linearity of the prism-dioptre method, the centrad system was introduced, but has never found much favour. In this

system 1 centrad is equal to a deviation of 1 unit at 100 units, but measured on an arc instead of a tangent scale. It will be seen that the increase of deviation increases linearly with the increasing value of the centrad deviation. The curved scale is difficult to construct in practice, but values can be calculated for a tangent scale in the appropriate way (Fig. 8.3).*

It will be remembered when dealing with refraction that the compound colours of white light are dispersed after refraction. This therefore also takes place when white light passes through an ophthalmic prism. This dispersion will obviously be more noticeable the greater the power of the prism, and if too great may become intolerable together with the distortion which occurs when looking through the prism obliquely. It is these factors which are accepted when the less thin, or ophthalmic, prism is used. No definite value of power can be assigned to 'thin' prisms as dispersion is relative. Prisms of a power less than 5 prism-dioptres are however generally accepted.

The amount of dispersion which takes place will vary not only according to the angle of incidence but also for differing types of glass. This means that if two rays of light are incident upon two separate prisms, having different refractive indices (n) and if the angle of incidence is the same in each case, the included angle between the blue and red rays of light after refraction will be greater in one instance than in the other (Figs 8.4 and 8.5).

*When measuring the power of a prism by systems (3) and (4) the prism must be held in the position of minimum deviation.

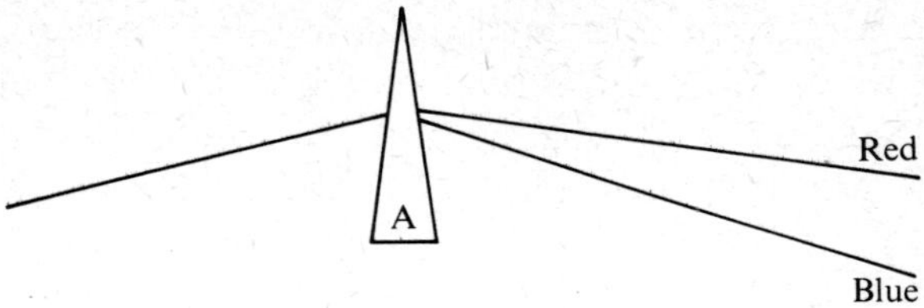

FIG. 8.4. Refractive index $(n) = x$.

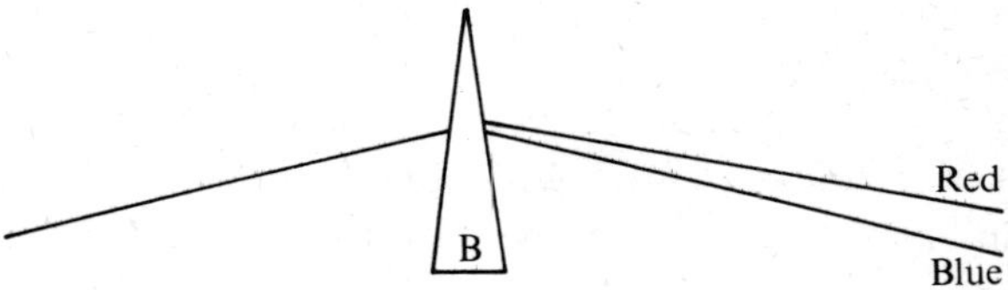

FIG. 8.5. Refractive index $(n) = y$.

It will be appreciated that if the prism producing the greater dispersion is made with a smaller angle, the angle of refraction would be reduced and the dispersion could be made equal to that of the first prism (Fig. 8.6). If these two prisms are now placed together with the base of one and the apex of the other coincident, the two dispersions will be almost cancelled out, but the incident ray of light will be re-

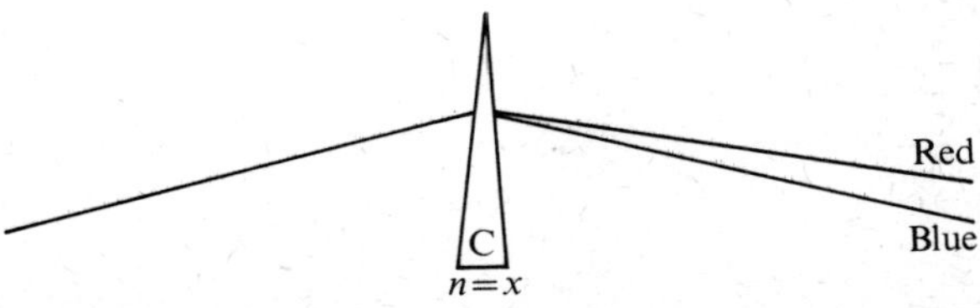

FIG. 8.6. Prism A with smaller angle dispersion equal to B.

fracted through an angle equal to the difference produced by the two prisms separately (Fig. 8.7).

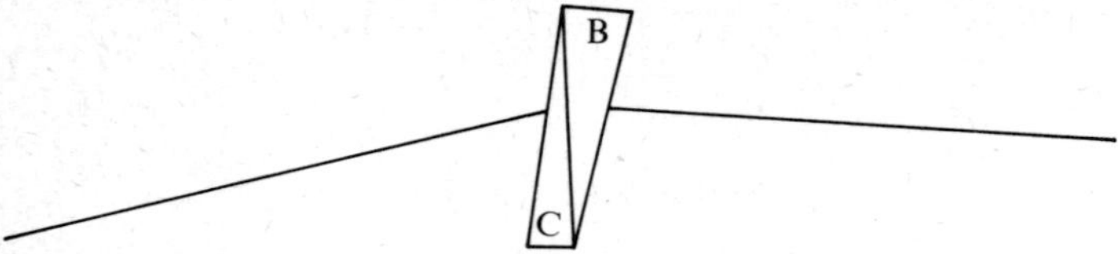

FIG. 8.7. The achromatic prism showing combined effect of prisms in Figs. 8.5 and 8.6.

This combination is referred to as achromatic, i.e. a prism producing deviation with little or no colour. Achromatic lenses are also made in a similar way using glass of two different refractive indices but in this case one lens would be positive and one negative to counteract the dispersions.

Further Reading

Fincham, W. H. A. and Freeman, M. H. (1974). *Optics* (8th edn). Butterworths, London.

PART II: VISUAL OPTICS

9 VISUAL ACUITY

Standard works have been established in ophthalmic optics for assessing patient's visual ability, and these define the patient's visual acuity. Methods of measuring visual acuity are numerous, but all have the same basis, namely the assessing of the patient's ability to differentiate between a series of lines and spaces representing some form, usually the shape of a letter. The basis of the standard for normal visual acuity is assessed largely from a calculated value of the size of the foveal cones. The foveal cones subtend an angle of approximately one minute of an arc. For two points to be differentiated simply at least one of two adjacent receptors must remain unstimulated. Objects or letters where bars and spaces subtend this angle at the retina of the patient are the minimum size that can be recognized. Standard vision is known as 1/1 or, more usually, 6/6 vision (the '6' refers to the standard distance (in metres) at which the chart is to be read). The form of the actual objects of letters used to test vision may vary cosiderably. One common basis for the so-called standard letter is to fit the letters into a square of 5 units by 5 units, each unit being the thickness of the letter, leaving a space of 1 unit between the limbs of the character formed, such as an 'E'. A common variation is to make the letter slightly taller than it is wide, this is found in most reading matter. Charts for measuring visual acuity consist of a series of lines of letters, where each line is half as big again as its predecessor.

Visual acuity is recorded as a fraction such as 6/9 or 6/12, where the denominator indicates the distance at which that particular letter would have to be placed to appear the same size as that of the normal 6/6 and the numerator the standard distance at which the chart is placed (in metres). Fig. 9.1 will

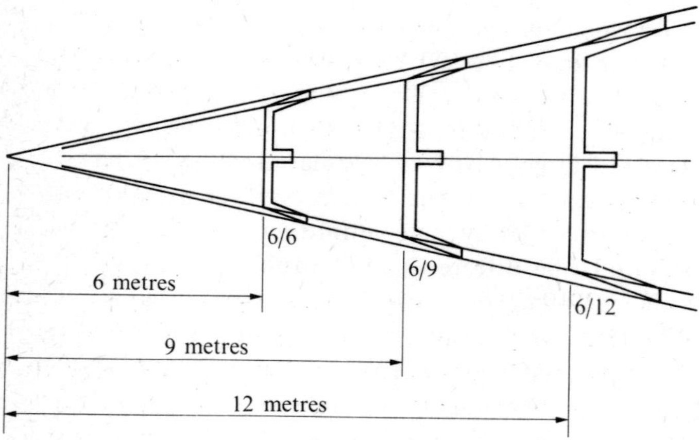

FIG. 9.1.

serve to show these visual acuity readings. Most charts will usually show two lines smaller than 6/6. The figure obtained for recording the visual acuity is subjective and therefore open to considerable variation. To minimize these, some standard form of illumination is also essential and characters on different charts must be printed in the same form. Difficulties of recording an accurate vision will arise when dealing with illiterate patients and also with young children. In the case of young children an accurate result may very often be obtained by checking the visual acuity on a number chart as many children will recognize numbers–having only to remember ten numbers as against 26 letters. For illiterate patients visual acuity can be obtained by using one single letter or character printed in a number of different positions for each line, the patient being asked to show which way up, or which way round, the letter appears to be.

For young children who do not know their letters, the Sheridan Gardiner test gives a practical and accurate assessment of visual acuity. The child holds a card on which a limited number of letters are printed. Letters are then selected by the examiner on a distance chart and the child finds the same letter on his card.

The opto-kinetic drum is valuable for assessing retinal response in babies. The opto-kinetic drum consists of a drum oscillating in a horizontal plane on which are printed dots of varying sizes which attract the child's attention. If they are visible the child's eyes will oscillate to follow the dots backwards and forwards. By decreasing the size of these dots a point can be found where the baby is no longer interested and from this a fairly accurate assessment of its visual acuity can be obtained.

Where space is limited charts may be situated next to the patient and viewed in a mirror so that the total distance of the chart to the patient via the mirror is still maintained at 6 metres. Where there is no obvious reason for a poor visual acuity in a patient, one should always suspect malingering. Various methods for detecting this have been put forward from time to time. Where only one eye is under suspicion, one of the simplest methods is to place a high plus sphere in front of the seeing eye and ask the patient to read with both eyes open. The malingerer may not suspect he is only able to use his 'poor' eye and will read with ease. The author has also had dramatic results in increased visual acuity by asking the patient to read the letters with the aid of 1/8 dioptre sphere, which the patient does not realize has little or no visual effect.

Further Reading

Emsley, H. H. (1976). *Visual optics* (5th edn), vol. 1. Butterworths, London.

10 THE EYE AS AN OPTICAL INSTRUMENT

In Chapter 7 on lenses Gauss's calculation was described for a series of lenses situated on the same optical axis on which the six cardinal points were defined. The eye, viewed as an optical instrument, consists of a number of interfaces, although not all the elements of this optical system are situated on a common axis. For the purpose of this chapter it will be assumed that the elements of the human eye are homocentric and consequently are represented diagramatically as in Fig. 10.1. The position, separation, and radii of curvature are shown on this diagram, but works on anatomy and visual optics should be referred to if more accurate dimensions are required.

From Fig. 10.1 it will be seen that the principle points and nodal points are extremely close together,

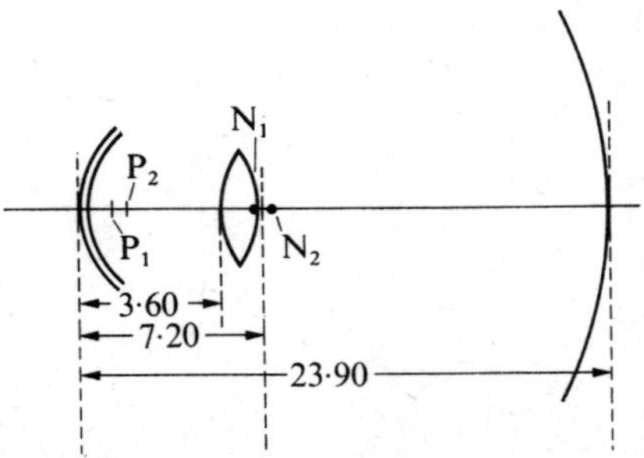

FIG. 10.1. Normal eye.

and for the purpose of greater simplification may be assumed to be single points producing Fig. 10.2,

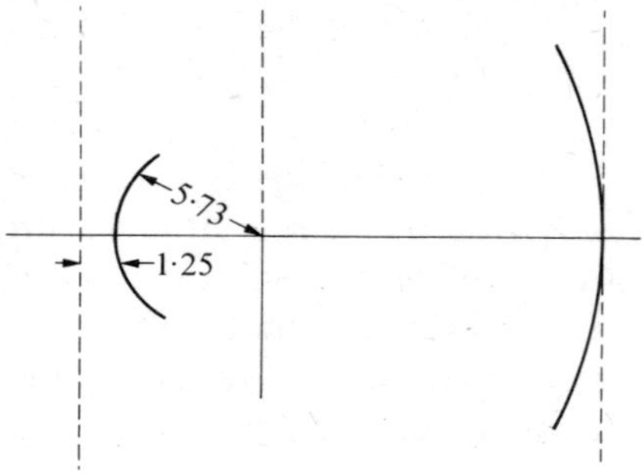

FIG. 10.2. Reduced eye.

representing a simplified optical system consisting of one refracting surface bounded by air on one side and water on the other. This latter diagram is referred to as the reduced eye.

Again using the diagram for the reduced eye, any calculation may be carried out for image size, the results being practically as accurate as would have been obtained taking into account all the elements of the optical system of the human eye. However, in one particular instance the reduced eye does not give accurate predictions, that of aphakia. Fig. 10.3 illustrates the dimensions of the aphakic eye, which are very similar to those of the reduced eye, with the lens removed. The cornea, however, is in the normal position as shown in Fig. 10.1.

It must also be remembered that the reduced eye applies only to the eye in its unaccommodated state,

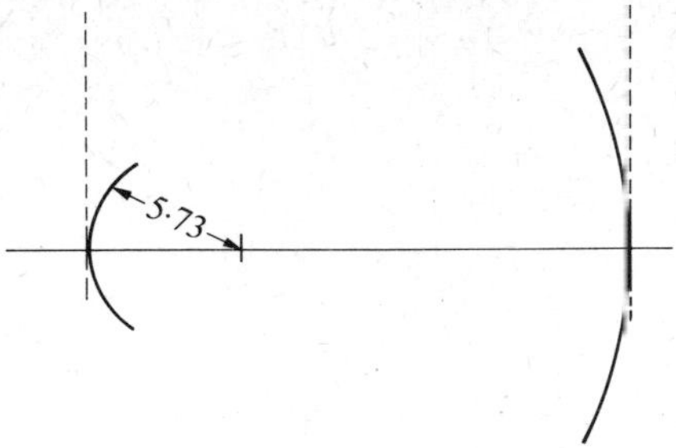

FIG. 10.3. Aphakic eye.

a different set of figures applying for every variation of accommodation.

In the foregoing chapter, it was assumed that all the elements of the optical system of the eye are centred on a common axis. This is not so because of displacement of the lens and/or apex of the cornea relative to one another. If a mean line which may be referred to as the optical axis, is taken from the posterior pole through the nodal point and apex of the cornea, it will be found that it is inclined at an angle to the visual axis (namely the line joining the point of fixation, the nodal point, and the macula).

The angle between these two lines is called angle alpha, having an average dimension of approximately 5°, and varying to some extent with the state of the ametropia. Angle alpha is normally positive with the fixation point lying on the nasal side of the optical axis, but sometimes may be negative. Where angle alpha is large, this result is taken into account in assessing the presence or otherwise of squints by corneal reflections.

A line drawn from the point of fixation to the centre of rotation is known as the fixation axis and makes a slightly different angle with the optical axis, referred to as angle gamma. (A further axis in connection with the optical system of the eye is sometimes mentioned, the so-called central pupillary line, and the angle between this line and the visual axis was referred to by Landolt as angle kappa. This angle is rarely specified, but further details may be obtained from any textbook on visual optics.)

The presence of refracting surfaces in the optical system of the eye gives rise to a number of reflections, the most obvious of which is from the anterior surface of the cornea. This is one of the first images looked at for a quick check for the position of the visual axis. Other reflections may be seen with care from the posterior surfaces of the cornea and anterior and posterior surfaces of the lens. The last three will only be of low brilliance as the relative refractive index between the two interfaces of the point of reflection are much closer than that of the anterior surface of the cornea. An absence of the reflection from the lens would be evident in a case of aphakia.

These reflections or images were first referred to by Purkinje are referred to as Purkinje images. The posterior surface of the lens, being concave to the incident light, produces an inverted image which may readily be seen if an object having distinguishable characteristics is viewed. Apart from the corneal reflections Purkinje images are difficult to see, not only because of their low brilliance, but also because of the all-defined bounded surfaces of the lens capsule, etc. The effect of accommodation on the curvature of the lens may readily be demonstrated by a change in the relative size of the third and fourth images when the eye is relaxed or accommodated. Excellent photographs by Fincham of the Purkinje images have also been published in books on visual optics.

Further Reading

Emsley, H. H. (1976). *Visual optics* (5th edn) Vol. 1. Butterworths, London.

11 AMETROPIA

Ametropia

Ametropia is a term that covers all the following refractive conditions: *hypermetropia* or long sight, *myopia* or short sight, and *astigmatism*, a refractive condition varying from one meridian to another which can be associated with hypermetropia or myopia.

The human eye consists of a complex series of optical media: the cornea, aqueous humour, lens, vitreous humour, and the retina on which the image falls. The simplified concept outlined in the previous chapter, the reduced eye, is therefore used to illustrate ametropia.

Before dealing with the ametropic conditions, the condition when no refractive error is considered to exist should be defined. An *emmetropic* eye is one which focuses any object situated at infinity (or any distance beyond about 6 metres), onto the retina, when the accommodation is at rest (Fig. 11.1). If

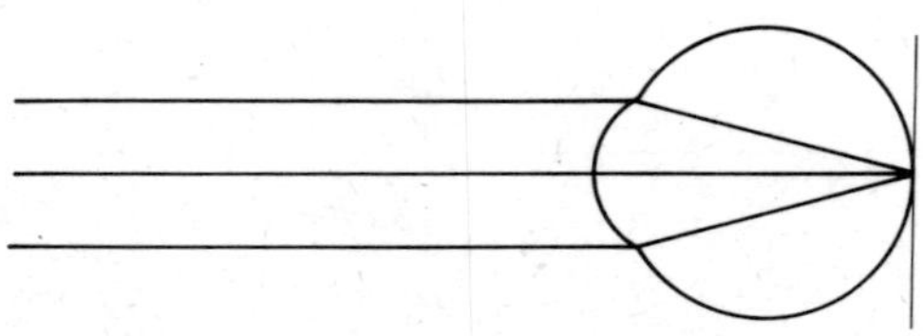

FIG. 11.1. Emmetropia.

the image falls in front of or behind the retina then one of the ametropic conditions is present.

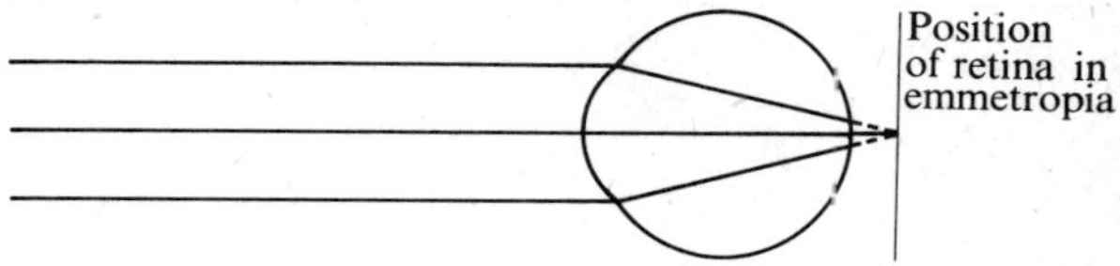

FIG. 11.2. Axial hypermetropia.

Hypermetropia

From Fig. 11.2 it will be seen that in the hypermetropic eye parallel rays of light would come to a focus behind the eye, forming a blurred image on the retina. This may be corrected by placing a positive spherical lens in front of the eye, converging the light before it enters the eye, so that it is subsequently focused on the retina by the optical system of the eye (Figs 11.3 and 11.4). Causes of hypermetropia can be due to a short axial length of the globe, giving rise to axial hypermetropia, or a low refractive power of the eye, giving rise to refractive hypermetropia, or a combination of both. Axial hypermetropia is more common.

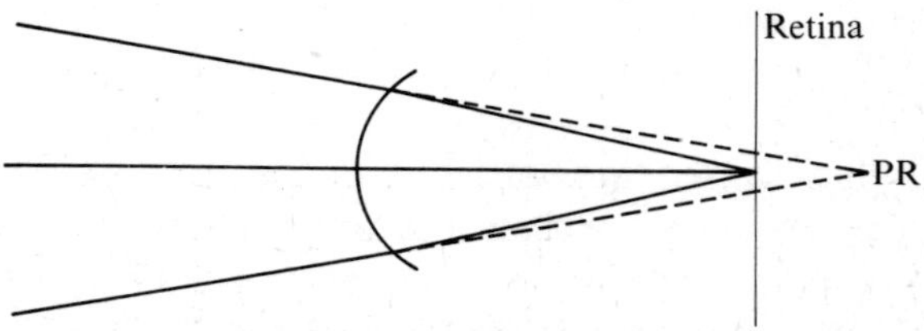

FIG. 11.3. Hypermetropic eye and correction.

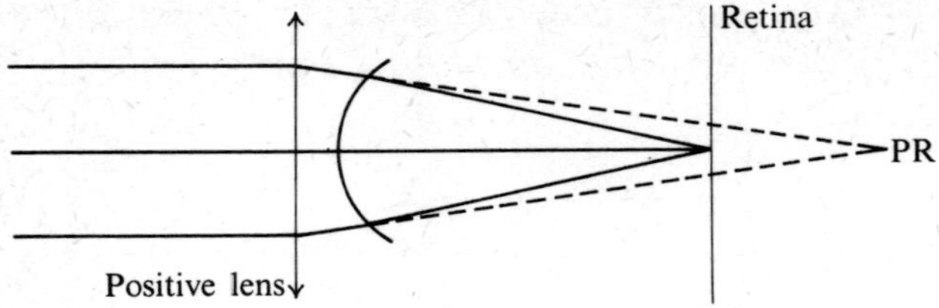

FIG. 11.4.

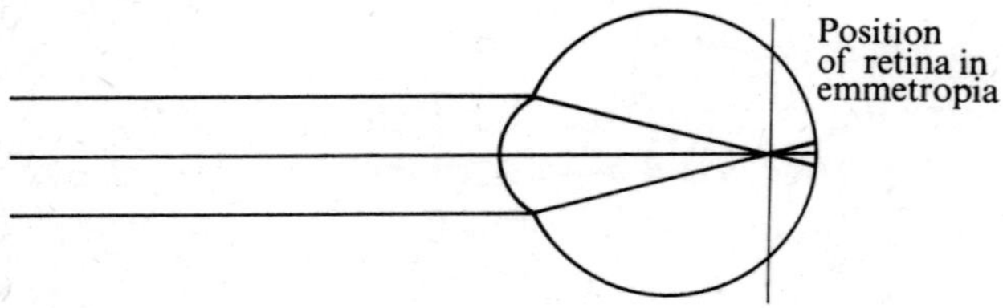

FIG. 11.5. Axial myopia.

Myopia

Fig. 11.5 shows the myopic eye and it will be seen that parallel rays of light from a distant object come to a focus in front of the retina, again producing a blurred image on the retina. A negative spherical lens in this case corrects the myopia, producing a divergence of light before it enters the eye, so that it is then correctly focused on the retina (Figs 11.6 and 11.7).

Similar causes are attributed for myopia as for hypermetropia, i.e. either the axial length being too long, in axial myopia, or the refractive error is too high, or there is again a combination of both. Axial myopia is the commonest condition.

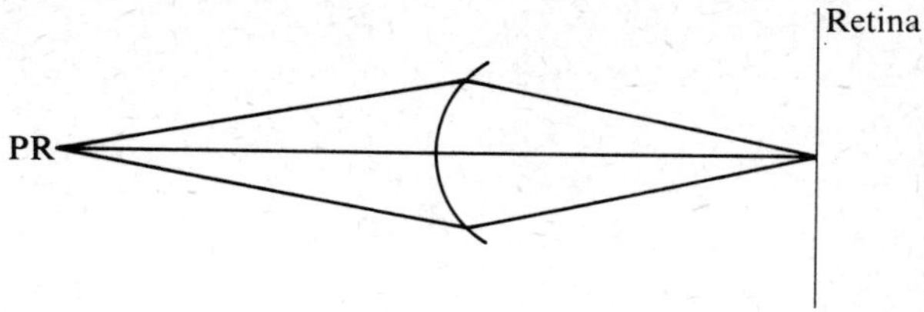

FIG. 11.6. Myopic eye and correction.

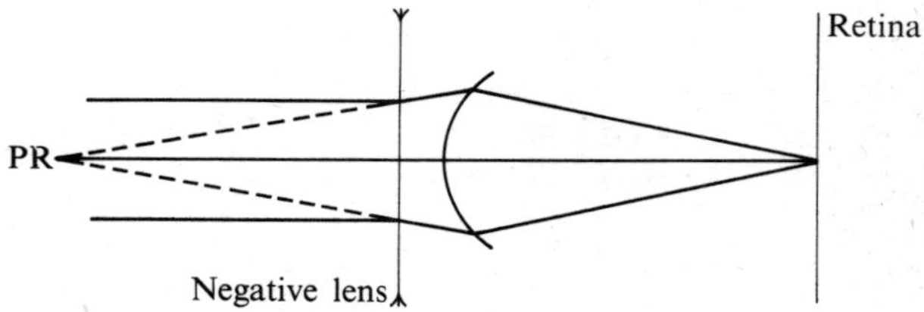

FIG. 11.7.

Astigmatism

In simple hypermetropia and myopia all meridians of the eye have the same power and are therefore corrected by spherical lenses. In astigmatism the meridians of the eye vary, usually in a regular manner, so that the maximum and the minimum meridians are at right-angles to one another, but not necessarily in a vertical or horizontal plane. This is referred to as regular astigmatism. The rays of light from a distant object do not come to a point focus in the eye but form two-line foci, as was seen in the construction of the diagram for a cylindrical lens (Fig. 7.16).

Where the meridional powers do not vary in a regular fashion, irregular astigmatism results. These cases are far less common and are generally associated with some pathological or traumatic condition.

The positions of the line foci relative to the retina give rise to fairly obvious categories of astigmatism:

(a) myopic or hypermetropic astigmatism (the required correcting lens is positive or negative in all meridians as the case may be),

(b) simple hypermetropic or myopic astigmatism (the correcting lens is plane in one meridian and plus or minus in the other meridian), and

(c) compound astigmatism (the correcting lens has one meridian positive and one meridian negative).

Astigmatism may arise from any distortion in the optical system, such as cornea or lens surfaces not being truly spherical, or the lens being suspended obliquely to the optical axis.

Optical principles of the correction of ametropia

In the hypermetropic eye, when the accommodation mechanism of the eye is relaxed, rays of light must be converged in order to be correctly focused on the retina. It will be seen from Fig. 11.4 that these rays of light come to a focus at point PR and it is therefore obvious that an object situated on the retina would appear, after refraction by the optical system of the eye, as an image at point PR. Point PR is known as the far point, or punctum remotum, being the furthest point away from the eye at which an object may be seen clearly, and in the case of hypermetropia must obviously be 'virtual'. Fig. 11.4 is similar to situation of the virtual image formed by a simple magnifying glass. In order for an object to be correctly focused on the retina, it is necessary for an image of that object to be formed at PR. In Fig. 11.4 the positive lens is focusing an image on the distant object at PR, which therefore creates the necessary convergence of light before entering the eye, obtaining a clear focus on the retina. The focal length of the lens required is therefore equal to the distance of the far point (PR) from the lens.

In the case of myopia the rays of light have to diverge, as will be seen from Fig. 11.6 before entering the eye. The rays diverge from a point in front of the eye. The far point in this case is now 'real' which is why a short-sighted person is unable to see clearly beyond a certain point, i.e. their far point. The negative lens required for correcting myopia is placed before the eye so that its focal point coincides with the far point, as does the positive lens in hypermetropia, and the necessary divergence of the rays of light from the distant object will have to be correct for the image to be correctly formed on the retina (Fig. 11.7). The focal length or power of the lens will then be equal to the distance of the lens from the far point.

It is important to note that the position of the correcting lens in front of the eye affects the required focal length for the correction of the ametropia. Bringing the lens closer to the eye would require a shorter focal length for the hypermetrope and a longer focal length for the myope. Where the lens is of low power this fact is of little importance, as the relative movement would be small in relation to the overall focal length. In lenses of high power the importance of the position of the lens before the patient will be greater and must be correctly adjusted. It is interesting to note that the movement of a 10 dioptre lens by 1 cm has the effect of altering the power of that lens by approximately 1 dioptre. Similar conditions apply when the eye is astigmatic, but here the optical principle involves the 'astigmatic' far points for myopia or hypermetropia according to the meridian in which the astigmatism is being investigated. The correcting lens is spherocylindrical and its focal length differs in the maximum and minimum power meridians, which again would coincide with the relative far points of the eye.

Presbyopia

Presbyopia is a condition arising from physiological changes within the eye, which become more apparent at the age of approximately 45. Assuming that the patient is either emmetropic or otherwise properly corrected with spectacles. The amplitude of accommodation decreases with age, and eventually the accommodation mechanism has to be used almost to its maximum extent, leaving little in reserve for focusing on a near object. This produces asthenopic symptoms (eye strain) and the patient is said to be presbyopic. It is at this stage that reading glasses have to be ordered, or stronger lenses for close work if a patient is already wearing glasses for distance correction. The amount of correction required will naturally vary but a fairly general rule is to relieve two-thirds of the patient's amplitude of accommodation. For example if a patient wishes to read at a distance of 25 cm, i.e. requires 4 dioptre of accommodation, and his accommodation is only 3 dioptre, two thirds would be relieved by 2 dioptre. The correct glasses for comfortable reading would be 2 dioptre strength. This is in addition to any necessary correction for distance vision.

In the uncorrected hypermetrope, who must exert his accommodation to see near objects clearly, the difficulty in doing close work is experienced earlier than in a corrected hypermetrope, as he has less accommodation in reserve. His hypermetropia will be found on examination and if he is prescribed correcting glasses he may not need additional power for close work. It would therefore be incorrect to term such a patient as presbyopic.

In the case of the myopic patient, the distance vision is normally already properly corrected. When only a small correction is necessary, the additional plus-power required to relieve the presbyopia could cancel this out, so that no glasses would be required for near work at all.

It can be seen, therefore, that the diagnosis of onset of presbyopia depends to a great extent on the refractive condition of the eye, particularly when it is uncorrected.

Anisometropia

Anisometropia is said to exist when the ametropia of the two eyes is different. Where this difference is small it is of no consequence, but large differences of 5 dioptre or more may lead to a disparity in the size of the retinal images, known as anisekonia, resulting in an inability of the brain to fuse the images.

This condition is more apparent when the ametropia is axial than when the ametropia is refractive. In the latter case, spectacle lenses should be fitted as near to the principle plane of the eye as possible, and although this is impracticable with normal spectacles, contact lenses very nearly approach this ideal. By far the greater number of refractive errors are axial, however, and this problem is not so easily overcome. Special 'size lenses', already mentioned in the chapter on lenses, can be tried but are apt to be rather bulky when produced. When the condition has obviously existed since birth, the author has always advocated a full correction for each eye at as early a stage as possible. In many cases this is fairly well tolerated and is preferable to relieving one eye incorrectly.

One further difficulty in connection with the correction of anisemetropia should be mentioned. When the patient looks directly through the centre of both lenses, no prismatic effect arises, but as the eyes are raised and lowered through this position, the prismatic correction before each eye varies according to the difference between the powers of the two lenses, giving the effect of a varying hyperphoria (see Chapter 16 Heterophoria). This may well create a variable diplopia, and lead to imbalance of the anisemetropic correction. The patient can overcome this difficulty himself to a certain extent, by restricting his eye movements to the central portion of the lenses. This difficulty may be well controlled by using contact lenses which remain centred.

The reasons for changes in the refractive error of the eye

In the early years of life the refractive error of the eye changes considerably, a baby normally being a hypermetrope of 4 dioptre which in a very large percentage of cases resolves into near emmetropia during childhood. Usually the refractive condition remains constant through adult life, and provided no illness occurs which would provoke a change, the prescription will probably vary very little, apart from the onset of presbyopia, until the age of 50–55 years. Diabetes can be a contributory cause to constant changes in the refractive error, and it is unwise to prescribe for a diabetic patient until satisfactory stabilization has been obtained. From the age of approximately 55 years sclerosing of all parts of the body, including the crystalline lens, starts to become apparent and can also alter the refractive condition by varying amounts in different individuals. This sclerosing may eventually lead to the lens becoming opaque requiring removal by surgery: this leaves the eye aphakic.

For a number of years before reaching the stage at which surgery is necessary it may be difficult to fit the patient satisfactorily with an optical correction. The hazy lens may give rise to considerable dazzle and glare in bright lights, not unlike the sun on a dirty window; and the irregular sclerosing may also produce a number of clear patches in the lens giving rise to monocular diplopia. Sudden changes in the refractive error are common during this time, the patient becoming less hypermetropic or more myopic. It is clearly very important to perform a careful examination of the fundus before the lens is too sclerosed to enable this to be done.

Aphakia

This condition normally follows operative procedure and the removal of the crystalline lens, and is not a form of ametropia. The overall refractive power of the optical system is considerably reduced, by approximately 10 dioptre; this means that an eye that was previously emmetropic becomes hypermetropic by 10 dioptre. The resulting hypermetropia in an aphakic eye will vary from 10 dioptre to a greater amount according to the previous hypermetropic state of the eye, or conversely by less than 10 dioptre if the eye was previously myopic.

The aphakic eye presents us with a totally different optical system, consisting only of the cornea bounding the air medium of the outside of the eye and the aqueous/vitreous medium of the inside of the eye, it is in effect equal to the reduced eye already explained (Chapter 10). It will be remembered that the nodal point is further forward in the reduced eye, resulting in an increased image size (aniseikonia). The correcting lens for the affected eye must of necessity be placed before the eye with some air space between the cornea and the correcting lens, which also increases the retinal image-size, introducing once again the difficulties of anisemetropia. The image-size disparity can be considerably reduced by the use of a contact lens for the aphakic eye, and usually results in reasonably satisfactory fusion of images.

Attempts have recently been made to replace surgically the crystalline lens with a lens made of an inert substance placed either in the anterior chamber or between the pupil and against the vitreous face. The physiological success of this procedure is good as the optical system is rendered near normal. Acrylic posterior lens implants had limited success, but anterior chamber implants appear to be more successful. After an operation for aphakia there is usually some image distortion owing to the incision and the subsequent suturing, and a varying degree of astigmatism is therefore usually found, in most cases the axis being horizontal. The amount of astigmatism has been considerably reduced following the use of improved sutures and operative techniques.

Further Reading

Emsley, H. H. (1976). *Visual optics* (5th edn) Vol. 1. Butterworths, London.

12 THE MEASUREMENT OF REFRACTIVE ERRORS

The methods of measuring refractive error are extremely varied, but in most cases a combination of methods coupled with one's own personal modifications is used.

For convenience only two main methods of measurement are to be described: objective and subjective refraction. These two terms describe the methods used; with the objective technique the refractionist uses specialist techniques to find the error, and with the subjective method he relies entirely upon the patient's observations. The method chosen will depend on the patient, his age, co-operation, and mental ability, and on whether other factors such as lenticular opacities, etc., are present.

For example, if the patient is a young child or elderly person who cannot co-operate, one would use only an objective method of refraction; conversely in an elderly patient with lenticular opacities objective refraction may be impossible and the subjective method would have to be relied upon. Method of measurement is discussed in the next sections.

Objective methods of refraction

There are two main methods of objective refraction, namely optometry and rectinoscopy.

Optometry

An optometer is an instrument which in its objective form measures the refractive state of the eye by viewing the image of a given object formed on the retina. The instrument will not be described here as it is rarely used. Although this is essentially an objective type of measurement, great co-operation is required from the patient and in common with all objective methods of measurement some form of final subjective test should where possible be carried out. Little is heard of subjective optometers although as recently as thirty years ago, these instruments were being sent through the post for patients to measure their own refractive errors. Recent advances in electronics have produced more sophisticated types of optometers; the latest, using computer techniques, has recently had its debut.

Retinoscopy

This is the method used by all ophthalmologists and opticians for the objective measurement of the refractive error. In the hands of a skilled operator it is extremely accurate and fast. Some co-operation from the patient is necessary, but the method used reduces this to a minimum. When active accommodation which cannot be controlled is encountered, cycloplegics may be used if necessary.

Briefly, the method consists of causing a moving object (a patch of light) to appear on the retina of the patient and with the aid of correcting lenses the image of this object is formed in the plane of the pupil of the examiner. The position of this image can be accurately judged by viewing the reflex of the moving object through the patient's pupil, taking into account its speed and direction of travel relative to the movement of the operator's retinoscope. The correcting lenses used will correctly focus the patient's retina in the same plane as the operator's pupil. The power of the lenses is then modified so that the patient's retina will be in focus with an object at infinity. The full description of the optical principle will be dealt with in a separate chapter.

Depth of focus

Depth of focus has briefly been mentioned in the chapter on lenses. In ophthalmology a small pupil will give rise to similar effects making retinoscopy difficult and the subjective measurement of refraction less critical, as larger changes of lenses will have to be made before blurring takes place. In some cases this can be an advantage, the small pupil producing a pin-hole effect, which may improve definition, but the loss of light through the small pupil may then well be a handicap. Difficulties associated with small pupils will be encountered particularly in patients being treated with miotics, i.e. for glaucoma. (Refer also to the pinhole test (Chapter 14).)

Further Reading

Emsley, H. H. (1976). *Visual optics* (5th edn) Vol. 2. Butterworths, London.

Giles, G. H. (1965). *Principles and practice of refraction and its allied subjects*. Butterworths, London.

13 THE OPTICAL PRINCIPLES OF RETINOSCOPY

The illuminating system

There are, broadly speaking, two different types of illuminating system used in retinoscopy: a self-contained hand instrument, consisting of its own lamp, optical system, and mirror, and a simple retinoscope mirror and convenient source of illumination, the latter usually being situated to one side and slightly behind the patient.

The former method is usually referred to as the luminous system of retinoscopy, and the second as the non-luminous, and of course the terms indicate only whether the instrument has the illuminating system contained *within it* or not. Modern self-luminous retinoscopes are today almost universally in use.

The optical system may best be described by referring to the non-luminous system. A source of light is placed beside the patient's head, the examiner usually being seated approximately 1 metre in front of the patient holding a small refracting mirror (Fig. 13.1). The image of the source of light will then be formed behind the examiner in the position shown, I. Any movement of the retinoscope mirror will now move this image in any given direction; thus if the mirror is tilted upwards, as seen in the diagram, the image will move downwards, so moving the sub-

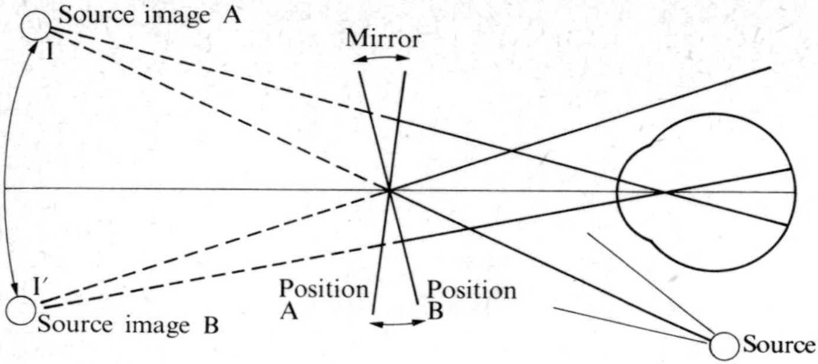

FIG. 13.1. Illuminating rays of light in retinoscopy.

sequent patch of light within the eye (which is in fact a further image of the light source) up. When the mirror image moves down, the light source image moves up and the patch of light in the eye down. It will thus be seen that the patch of light within the eye will move in the same direction as the mirror. The refractive condition of the patient will not in any way affect this direction of movement, apart from the speed at which it moves. In order that the examiner may view directly along the path of the rays of light entering the patient's eye, the retinoscope mirror has a small hole in the centre.

In the case of a self-contained instrument a virtual image of a light source is formed by the lamp within the focal length of a spherical lens focusing the rays of light being reflected by a similar mirror, as is used in the method described, but smaller in size and integral within the instrument. Fig. 13.2 shows the skeleton optical system of such a retinoscope.

Having established that we control the movement

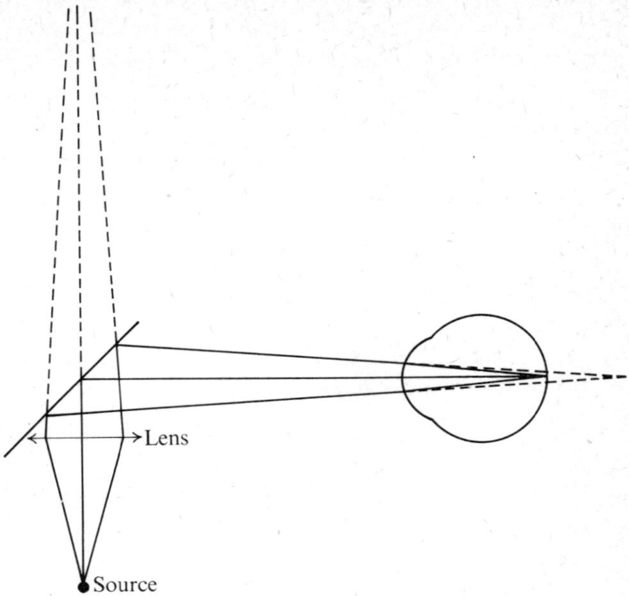

FIG. 13.2. Self luminous retinoscope.

of a patch of light focused on the retina, we can now examine the optical system used for assessing the refractive error of the eye.

The viewing system

It may now be assumed that the moving patch of light on the retina is an object, which will be imaged by the optical system of the eye, between the eye and infinity if the eye is myopic, or behind the eye if it is hypermetropic. This may easily be understood by referring to Chapter 11 on ametropia, and in particular the discussion of the position of the far point in hypermetropia and myopia.

We will take first the case of the myopic eye where the far point is situated between the examiner and the patient and, for the sake of simplicity, we will also assume that the examiner and patient are 100 cm apart.

If the patient is 4 dioptre myopic, the far point will be at position 1, 25 cm from the patient, and the moving image of the patch of light will move from position A′ to B′ (Fig. 13.3). Thus this movement will be in the opposite direction to the movement of the patch of light (from A to B), and also, of course, to that of the mirror (this is termed an 'against' movement). If the patient is now rendered less myopic by the use of minus lenses, the far point will move nearer the examiner, to positions A″ and B″. The image will still move in the opposite direction,

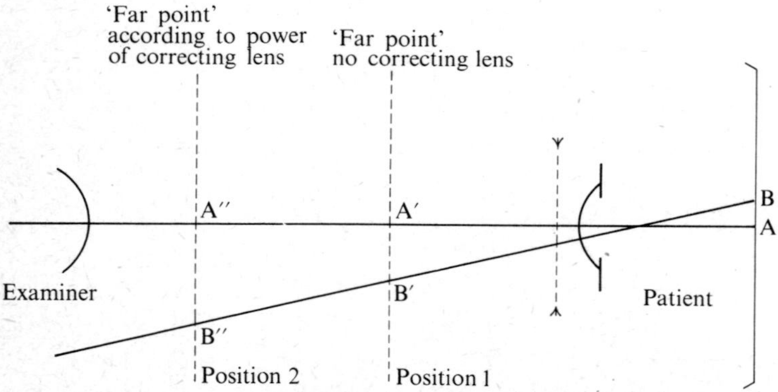

FIG. 13.3. Viewing rays of light in retinoscopy.

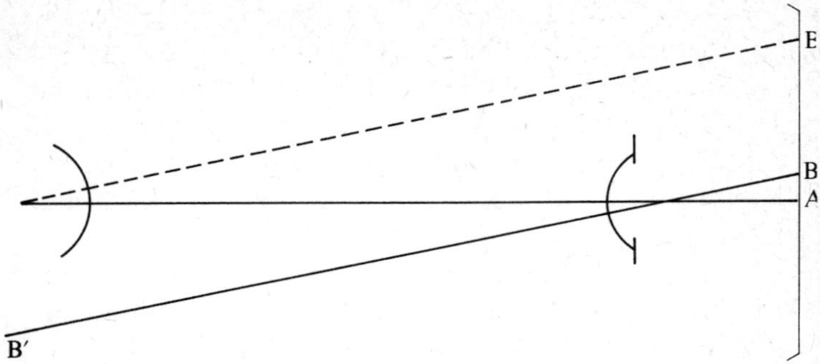

FIG. 13.4.

but if the movement of the patch of light in the eye is kept constant, it is clear that the nearer the far point is to the examiner the faster the image in the far-point plane will have to move, having a great distance to cover in the same time: this is readily observed when carrying out retinoscopy. If the far point is now made coincident with the examiner, the patch of light will move very rapidly and will only momentarily be visible, when it coincides with the examiner's pupil. This is called the point of reversal. By further reducing the myopia of the patient by means of lenses, the far point may be formed behind the examiner. As he will now no longer be able to view the real image in the far-point plane, he must focus the rays of light entering his eye, which will appear to be coming from a point situated along a line parallel to the rays of light forming the image at the far point. It will be seen from Fig. 13.4 that an upward movement of the patch of light on the retina will now appear as an upward movement of the image as seen by the examiner, or a 'with' movement. It must be realized that all these observations are made within a very narrow cone of light emerging from the patient's pupil, and the only visible rays are those passing from the patient's pupil into the examiner's pupil, which, of course, is an extremely confined area.

It is obvious that the critical point in these observations is the point of reversal, as it is at this point that the patient's far point may be accurately positioned. It may also be readily seen that no matter what the patient's refraction is, by placing plus or minus lenses in front of the patient's eye, the far point of the combined eye and lenses can always be made to coincide with the pupil of the examiner. No separate diagram has therefore been shown to illustrate the point of reversal in hypermetropia. If the accommodation is relaxed this will mean that the patient is accurately corrected for the distance at which the examiner is working, say 100 cm, and the lenses being used to create this circumstance must be reduced in power, algebraically, by 1 dioptre to obtain the patient's correct distance-refraction. Where the examiner works at a distance less than 100 cm a suitable correction must be made, say 1.5 or 2 dioptre.

The image at the point of reversal can be extremely distorted and one normally tries to assess this critical point by using a lens of low power which will change an against movement to a with movement, or vice versa. The correct lens for the point of reversal would then be of half this power plus all the checking lenses already in front of the patient.

It will thus be seen that an 'against' movement in retinoscopy indicates a myopic patient where the refractive power is greater than the 'dioptric' distance at which the examiner is working, and a 'with' movement indicates grades of hypermetropia and myopia of a value less than the working distance.

In some instances the retinoscope in use may be adjusted to produce a real image, instead of a virtual image, between the examiner and the patient. This may also be achieved by using a concave mirror and a distant light source. When this illumination system is used, the movement of the patch of light on the retina will be in the opposite direction to that of the movement of the mirror, so that all reflex movements relative to the mirror will be exactly opposite to those described above. Most modern instruments are capable of being focused to produce this effect.

The external light source is rarely used today as there is so much scattered light in the examining room, requiring a much stronger illumination to start with, which in the case of the self-contained hand instrument may be readily controlled by the examiner. Some examiners prefer using concave (real) image retinoscopes in high myopia but the author has never found this an advantage.

Further Reading

Emsley, H. H. (1976). *Visual optics* (5th edn) Vol. 1. Butterworths, London.

Giles, G. H. (1965). *Principles and practice of refraction and its allied subjects*. Butterworths, London.

14 SUBJECTIVE METHODS OF REFRACTION

As this name implies, subjective refraction relies mainly on the subjective reaction of the patient. Two of the most common methods will be described, the second method being almost universally used today.

The fogging method

For the fogging method, plus or minus lenses are placed in front of the patient's eye in such a way as to ensure that all rays of light entering the eye come to a focus just in front of the retina. This will produce a slightly blurred image of a distant object, i.e. the test chart, and, where astigmatism is present, a greater blurring in one direction than in the other. This difference in blurring may then be corrected with cylindrical lenses, the final correction being made with spherical lenses to produce an accurate focus of the distant object on the retina. The method usually adopted to achieve this end is with the aid of spherical lenses only, a series of lenses being tried until the best possible vision is obtained with maximum plus or minimum minus spherical power. If no astigmatism were present these lenses would represent the true refractive error. Plus spherical power is then added until the test chart is blurred to approximately 6/18 or 6/24, which, assuming the astigmatism, if present, is not very high, would bring both line foci in front of the retina. Where astigmatism is extremely marked, the measurement of the cylinder power is more difficult, but a high cylinder power would normally be apparent and indicated by a very low visual acuity when trying the best spherical correction.

Assuming the two line foci are made to form just in front of the retina, the patient is then asked to view a series of radial lines, usually placed 10° apart, and asked to state in which meridian these lines appear clearest, or alternatively the most blurred. The clearest lines will be nearest to the retina but will in fact be blurred along their length, and the lines at right-angles to them will appear most blurred, being blurred across their width. To make these latter lines clear it will be necessary to place a negative cylindrical lens with its power along the direction in which the lines are blurred, i.e. with its axis parallel to the direction of the lines of the blurred meridian. Cylindrical lenses are therefore placed in the trial frame corresponding to the axis direction, their power being altered until all the radial lines appear to be equally blurred.

The point at which this equal blurring occurs can be extremely difficult to assess, and a block of lines, having a series of parallel lines, one half set at right-angles to the other, often has to be used. This method is commonly referred to as the fan and block method for assessing astigmatism. This block is rotated until one set of lines lies in the blurred meridian, and the cylindrical lenses altered until both halves of the block by direct comparison appear to be equally blurred. The spherical correction in the trial frame may then be modified to obtain the best visual acuity, bearing in mind the general rule of maximum plus and minimum minus to keep the accommodation as relaxed as far as possible.

Many different methods and charts have been devised for assessing the amount of astigmatism in this way, but all are based on the same principle, varying in an endeavour to make it easier for the patient not only in matching the sets of lines, but also the position of the clearest and most blurred meridians. The complete routine is carried out on both eyes, and any final adjustments or any other vision tests are made binocularly.

Cross-cylinder method of refraction

This method is probably the most universally practised method of subjective refraction for assessment of astigmatism. The method is quick and accurate and does not rely as much on the patient's subjective responses. The best spherical correction is first obtained in the way described for the fogging method, and the patient left very slightly undercorrected, i.e. approximately −0.25 added to the best spherical correction found. This makes the patient exert the small amount of accommodation which is thought necessary for this method of refraction. At this point, when astigmatism is present, one line focus will be situated in front of the retina and one behind. Any two cylindrical lenses of equal but opposite powers, with their axes set at right-angles to each other, are now placed in front of the eye with their axes in line with the maximum and minimum meridians of refraction. It will readily be appreciated that these two lenses will either make the two line foci further apart, by adding to one in a positive direction and to the other in a minus direction, or, if turned through 90° will have exactly the opposite effect and will bring the two line foci together. If the patient is viewing a letter chart, the letters will appear much clearer in one position than in the other.

The two cylinder lenses are made in the form of a single lens mounted in a ring with a handle set at 45° to the axis of the plus and minus cylinder. The apparatus, known as a crossed cylinder, is available in various powers, usually of 0.25, 0.50, and 1.00 dioptre, with the axes of the two cylinders being set at 45° to the handle of the cross cylinder. The change from the plus to minus meridians can be simply achieved by twisting the instrument with the fingers along the axis of the handle. This produces the same effect as rotation of the cylinders about their optical axes, but is easier to achieve. A minus cylinder is then placed in the trial frame with its axis parallel to that of the minus cross cylinder when held in the clearest position. This test may then be repeated until the letters appear to be equally blurred when the axes of the cross cylinder are turned through 90°, which will indicate that in one position the two line foci will be situated in front of and behind the retina, but in the second position their positions will be reversed, i.e. the one in front will now be behind and vice versa.

The foregoing has assumed that the axes of the

astigmatic meridians are known, and they would in most cases be obvious from the retinoscopy technique. In the few cases where retinoscopy cannot be carried out, the axis of the cylindrical lens may be approximated by revolving the whole cross-cylinder unit until one clearer position is found and the foregoing procedure carried out from that position.

The accurate setting of the axis must then be checked. This may again be achieved by using the cross-cylinder unit. When two cylindrical lenses of the same sign are placed one on top of the other with their axes not parallel the resultant effect will be a lens of a different power (spherical if both lenses are equal and the axes at right-angles). If the lenses are not at right-angles the resultant will always be a cylindrical element of different power, the axes of which will be nearer that of the cylinder with the higher power. It is the resultant rotating of the axis in this way which is used for assessing the axis position of the lens before the patient. The cross-cylinder lens is held in front of the cylindrical lens with the handle in line with the trial lens axis. The combined effect of the cylinders will result in the axis moving towards the cross-cylinder axis of the same sign, as outlined above. Twisting the cross-cylinder will move the axis in the opposite direction. If either of these positions is clearer, a movement of the cylinder axis in the direction of clearest position should be made, always towards the plus direction for plus cylinder and minus for minus. When both positions are equally clear or blurred the position of the trial lens is correct. The cylindrical and spherical correction should be verified every time a change of cylindrical power is made. A final check is then made of the spherical power.

The two foregoing methods of subjective refraction are by no means the only methods commonly in use. A discussion of this subject would not be complete without a brief reference to the *duochrome test*, which is now in fairly regular use in one form or another. The duochrome test makes use of the property of chromatic aberration of the eye, which means that the eye has a shorter focal length in the green–blue end of the spectrum and a longer focal length in the red end of the spectrum. If a series of red and green letters are presented to the patient, the letters of that colour which is focused nearest to the retina will appear clearer. The other letters may then be made clearer by adding the appropriate plus or minus lenses, that is minus to bring the green letters clearer and plus to bring the red letters clearer, the optimum position being where both sets of letters are clear and thus situated equidistant in front and behind the retina. Opinions of the value of the duochrome test vary between individual practitioners.

The pin-hole disc

Where a satisfactory visual acuity cannot be obtained the value of the information gained by the use of a pin-hole disc must not be forgotten. The image on the retina formed by the rays of light passing through the pin-hole disc may be considerably clearer as only a very narrow pencil of rays will now be in use. If a considerable increase in vision is obtained with the use of a pin-hole disc, it may normally be assumed that there is no gross pathology of the retina, certainly at the macula, and the lack of good visual acuity can therefore be attributed to incorrect refraction or some modification of the optical system, i.e. lenticular opacities, and irregular astigmatism.

Further Reading

Emsley, H. H. (1976). *Visual optics* (5th edn) Vol. 1. Butterworths, London.

Giles, G. H. (1965). *Principles and practice of refraction and its allied subjects*. Butterworths, London.

15 OPHTHALMOSCOPY

Whenever a patient attends for consultation, a routine examination of the exterior and interior of the eye and lids must be carried out. Most of the exterior examination of the cornea and anterior chamber can be carried out with the aid of a torch and magnifier in conjunction with a slit lamp and microscope. The examination consists simply of illuminating the eye with a special optical system and viewing it minutely under high magnification with a binocular microscope. When the deeper layers and the interior of the eye have to be examined, the optical system of the eye has to be taken into account and a different technique is used; this is referred to as ophthalmoscopy. Ophthalmoscopy is not confined solely to the examination of the retina; the optical system of the ophthalmoscope may be adjusted according to the position of the particular part of the eye being examined.

Two systems of ophthalmoscopy are available, which use different optical principles. Using direct ophthalmoscopy, the media etc., are viewed directly, and this method gives a relatively high degree of magnification. Using indirect ophthalmoscopy an image of the retina is formed with a lower magnification but a wider field of vision. Both methods may be used during an examination, the indirect method is advantageous, if not essential, when viewing retinal detachments and where a wide field is required. The direct method has the advantage of the higher magnification.

The ophthalmoscope

The simplest form of ophthalmoscope is a plain-mirror retinoscope using some form of illumination in a similar manner as in retinoscopy. Using this mirror the general appearance of the cornea and lens may readily be observed by reflecting light into the patient's eye, with the examiner sitting approximately 50 cm away. Any opacity of cornea or lens may easily be seen against the retinal reflex. The apparent position of this opacity will automatically be related to the pupil of the eye; if the examiner's head is moved from side to side an opacity situated in the plane of the pupil will remain stationary whereas an opacity situated in the cornea will appear to move across the pupil in the opposite direction to any movement of the examiner. Conversely any opacity situated in the lens behind the pupil will appear to move in the same direction as the movement of the examiner. With this simple instrument, therefore, the position of any opacity in the interior of the globe may be fairly accurately fixed before any further examination is carried out. To examine more closely the examiner must be nearer to the patient. This requires a strong positive lens in the viewing system and adequate illumination may be difficult to obtain. Special instruments are therefore made to facilitate close-up ophthalmoscopic examinations.

Optical principles of ophthalmoscopy

Direct ophthalmoscopy

When the examiner views the retina of the eye all rays of light emerging from the ophthalmoscope will pass through the optical system of the patient's eye, and if he is emmetropic they will emerge parallel. If, therefore, we illuminate the interior of an emmetropic eye by some optical system, a real inverted image of the fundus will be formed at infinity. If now the examiner places his eye, which we will also assume to be emmetropic, in the path of these parallel rays of light, he will in fact receive a real image on his own retina of the retina of the patient. This quite simply is the viewing system of any direct ophthalmoscope. Where a patient is not emmetropic, a suitable lens will have to be placed between the examiner and the patient to counteract the ametropia, as also would be necessary should the examiner not be wearing his spectacles. A combined correction of the ametropia of the patient and the examiner may of course be accomplished in one single lens, and most ophthalmoscopes are therefore designed with a series of plus or minus lenses which may be inserted in the sight hole of the instrument.

The illuminating system of the modern direct ophthalmoscope is incorporated within the instrument, as in a self-luminous retinoscope. The optical system which is often employed is shown in Fig. 15.1. The light from lamp L passes through a condenser C and projector lens P to a focus beneath the sight hole of the mirror N. After reflection, these rays of light will then diverge, being refracted by the optical

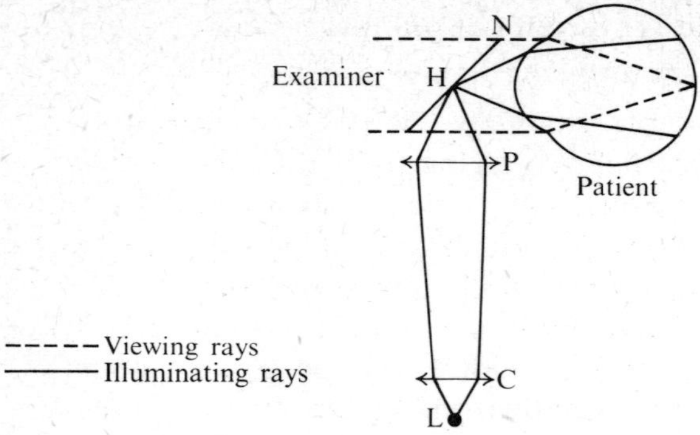

FIG. 15.1. Direct ophthalmoscope.

system of the patient's eye, forming an evenly illuminated patch on the retina. The optical properties of this system will vary according to individual manufacturers but the principle remains the same.

As the illuminating and viewing system must of necessity be almost coincident, reflections from the cornea are distracting, but these can be reduced to the minimum by careful design of the optical system. The size of the patient's pupil will affect the ease with which the instrument may be used; the larger the pupil the easier it is to avoid the corneal reflections, and also the greater the field of vision. A further factor influencing the efficiency of the ophthalmoscope is the size of the sight hole, which again if too small will reduce the field and the amount of light passing to the examiner, but on the other hand will tend to reduce the corneal reflections. Here again the manufacturer endeavours to fix an optimum size. A further factor affecting the field of vision is the distance of the examiner from the patient. The ophthalmoscope is made with a minimum protrusion, enabling the examiner to get as close as possible to the patient. With direct ophthalmoscopy the fundus examined is viewed, when the eye is emmetropic, under a magnification of approximately 15 times, this being calculated in the similar way as the magnification of a simple magnifying glass. The magnification is equal to:

$$\frac{250\text{ mm (distance of distant vision)}}{16.6\text{ mm (focal length of eye)}}$$

As the focal length in a myopic eye is shorter and a hypermetropic eye is longer, the magnification will be greater or less according to the ametropia.

Indirect ophthalmoscopy

When using direct ophthalmoscopy the fundus is viewed, as the name of the system implies, directly, without any intermediate image being formed. When the indirect system is used, a real image of the fundus is formed, and this is viewed by the examiner. This image is formed by placing a strong positive lens between the patient and the examiner. The fundus is suitably illuminated, by an illuminating system which is described later. The examiner and patient sit approximately 60 cm apart and (Fig. 15.2), a strong positive lens C2 of approximately 13–20 dioptre, and usually specially ground to reduce distortion, is held at approximately 9 cm from the patient's eye. The lens is held in such a position that an image of the patient's pupil is formed at the eye of the

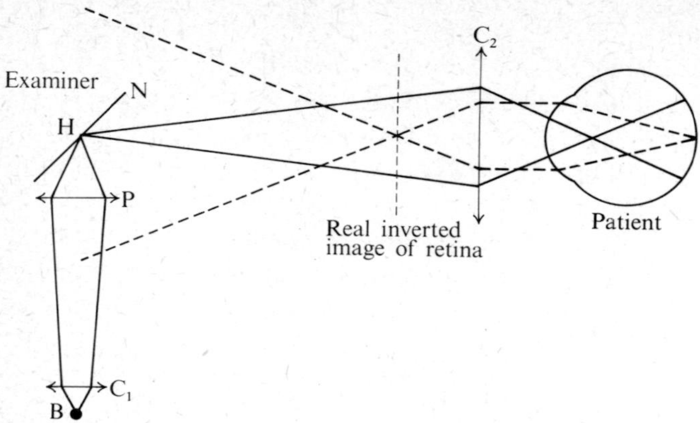

FIG. 15.2. Indirect ophthalmoscope.

examiner. At this point the aperture of the positive lens will appear to be filled with a view of the fundus, which has in fact formed a real image at the focal point of this lens. When the patient is emetropic, the magnification under these circumstances will be approximately five times, and may be larger or smaller in hypermetropia or myopia according to the positioning of the lens. A consequently larger field of vision than indirect ophthalmoscopy will therefore be obtained.

The indirect system of ophthalmoscopy is particularly useful in cases of high myopia, when a small field of vision and a large magnification makes direct ophthalmoscopy particularly difficult. A similar, but modified system of the illumination employed in direct ophthalmoscopy is used but as the distance from the patient is greater a stronger lamp is required. This is usually supplied from a low-voltage mains supply. The strong positive lens is used not only for forming the indirect image but also in the illuminating system. Some clinical ophthalmoscopes can be modified to form indirect ophthalmoscopes. The complete optical system of indirect ophthalmoscopy is shown in Fig. 15.2.

An image of the bulb B is formed by the lenses, again close to the sight hole H. This image is then again focused by the high-power lens C2 in the plane of the pupil forming a wide angle cone of light illuminating the retina. Rays of light leaving the retina will emerge from the eye parallel (in emmetropia), forming the inverted fundus image at the focal point of lens C2. This image is then viewed by the examiner, usually with a low-power positive lens to correct for the position of the image and to avoid using the examiner's accommodation. Two extra reflections are produced, one from the front and one from the back surfaces of the positive lens, which may be easily avoided by slightly tilting the lens about an axis at right-angles to the optical axis of the system.

All direct ophthalmoscopes are very similar, with different refinements and modifications introduced by various manufacturers. A number of different indirect ophthalmoscopes are produced, ranging from simple hand-held instruments, as described, to large stand instruments, some of which incorporate binocular eyepieces for viewing the fundus stereoscopically. In some types of indirect ophthalmoscopes, the illumination and binocular eyepiece unit are fixed to a head band, enabling the examiner to have complete freedom of movement and to examine the fundus from all positions.

Further Reading

Emsley, H. H. (1952). *Visual optics*, Vol. 1. Butterworths, London.

16 HETEROPHORIA

It would be inappropriate to deal fully with heterophoria in a book of this type but it is felt that some mention of the optical principles of the measurement of heterophoria, and a brief resumé of their relationship to the refractive error, is essential. The published works dealing with the physiology of the extra-ocular muscles should be consulted for further information.

Heterophoria is a general term used to indicate a condition in which both the visual axes are directed towards the fixation point but deviate on disassociation, i.e. there is a latent deviation which is normally held in check by the power of fusion. The easiest way in which disassociation may be accomplished is to occlude one eye, which will then, in theory, return to its position of physiological rest.

The extent of heterophoria is determined by measuring the angle between the two optical axes. One of the simplest methods uses what is termed the Maddox rod. The Maddox rod consists of a very high-power cylindrical lens, usually coloured, mounted in a standard trial case rim. This is placed before one of the patient's eyes and he is asked to view a small spot of light, either at the distant or near vision position. The eye not covered by the Maddox rod will view directly the spot of light along the visual axis. The visual axis of the other eye will move to the position corresponding to physiological rest, and the patient will see a coloured streak of light at right-angles to the axis of the Maddox rod. This may appear to be situated to the left, right, above, below or on the distant spot of light according to the position relative to the macula that the image formed by the Maddox rod falls.

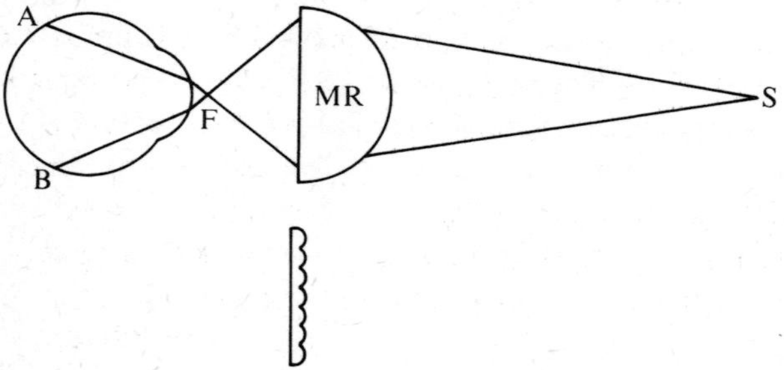

FIG. 16.1. Maddox rod.

The formation of this image is as follows, referring to Fig. 16.1: A point source of light S is imaged by the Maddox Rod MR (the image is a cross-section at right angles to the axis of the cylinder) at point S. The strongly divergent rays of light will be slightly converged by the eye but will form a blurred image on the retina extending from A to B. In the plane at right-angles to that of Fig. 16.1 the rays of light will not be converged, but will pass into the eye where they may be brought to the focus at point C (Fig. 16.2). The combined effect of Figs. 16.1 and 16.2 is to produce a streak of light on the retina at right-

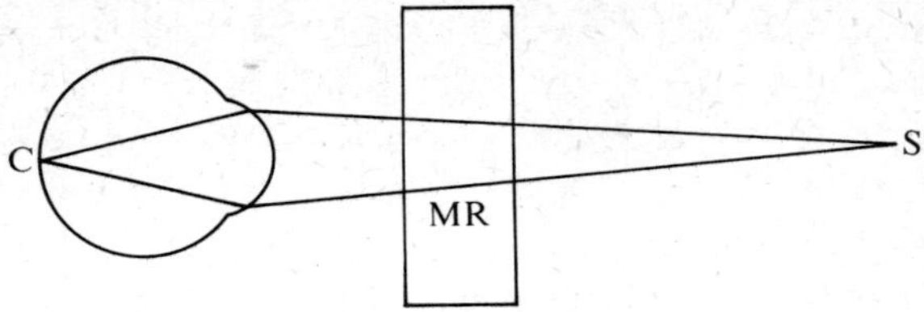

FIG. 16.2.

angles to the axis of the Maddox rod, in much the same way as a line focus is formed by a cylindrical lens. In order to make sure the image produced by the Maddox rod is long enough, it is usually made up in a series of cylindrical lenses placed one beside the other. The images will then overlap and optically the same description will apply for each one.

There are numerous other methods of determining the degree of heterophoria, employing different systems of disassociation, but the Maddox rod is probably one of the most commonly used.

Briefly the terminology used for the deviation of the eyes from the 'normal' or orthophoric positions are:

(1) esophoria, where the eyes converge on disassociation,

(2) exophoria, where the eyes diverge on disassociation, and

(3) hyperphoria for a vertical disassociation.

When the eyes show a constant deviation the patient will have an eso/exotropia – a squint in lay terminology – requiring specialized investigation and treatment.

Heterophoria

Only a brief description will be given here of the relationship between the refractive error and heterophoria. It must be stressed, however, that a complete investigation of the action of the extra-ocular muscles must be carried out if any patient is heterophoric.

In an uncorrected hypermetrope some convergence must be necessary to focus distant objects giving rise eventually to an esophoria and asthenopic symptoms in maintaining single vision. This would probably be considerably relieved by a prescription. If, however, after obtaining the fullest possible refraction, a large degree of esophoria still exists it may be wise to try the effect of an increased positive lens.

Where a patient is myopic with esophoria, special investigations must be carried out; a minus lens tends to aggravate the condition.

Exophoria, requiring a convergence effort to overcome it, is more easily carried out and probably gives rise to a smaller number of asthenopic symptoms. Where the patient is myopic, the fullest myopic correction will probably be more readily tolerated.

Hyperphona requires special investigation, and it may be necessary to prescribe a vertical prism. The treatment is beyond the intended scope of this book.

The subject of oculo-motor and accommodative balance is a highly complex subject which, in conjunction with the diagnosis and management of squint, calls for an entire bibliography of its own.

Further Reading

Burian, H. and Von Noorden, G. K. (1974). *Binocular vision and ocular motility*. C. V. Mosby, St. Louis, Mo.

Cashell, G. T. W. and Durran, I. M. (1974). *Handbook of orthoptic principles* (3rd edn). Churchill Livingstone, Edinburgh.
Duke-Elder, S. (1973). *Systems of ophthalmology*, Vol. 6. Henry Kimpton, London.
Emsley, H. H. (1952). *Visual optics*, Vol. 2. Butterworths, London.
Von Noorden, G. K. and Maumenee, A. E. (1967). *Atlas of strabismus*. C. V. Mosby, St. Louis, Mo.

PART III: ALLIED PRINCIPLES

17 THE FITTING OF SPECTACLES

When a patient has been refracted and glasses prescribed, it is of prime importance that the lenses should fit accurately on the patient's face. When lenses were of the round type, it was relatively easy to check that the spectacles had been fitted with their geometrical centres coinciding with the optical axes of the patient's eyes, either for distance or for near vision. Where the geometrical centres do not coincide this may be because the glasses have been made an irregular shape for cosmetic appearances, in which case the optical centres of the lenses should have been decentred inwards or outwards, to make the glasses optically perfect. With the modern trend in spectacles towards irregularly-shaped lenses, the geometrical centre is sometimes almost impossible to define, and the optical centre must therefore be marked before the accuracy of fitting can be checked. Any inaccuracy in the vertical meridian of the centre will probably give rise to greater discomfort, as this may introduce a vertical phoria which the patient may not be able to overcome. The glasses should therefore be carefully checked for correct fitting. Any misalignment in the plane of the lenses will give the effect of turning the lenses about their optical axes, again introducing serious errors where an astigmatic correction has been prescribed. The tilting of the frame may also be critical, particularly in high-power lenses, and a mean position is normally fixed such that the patient will look through the lenses with his optical axes slightly depressed and normal to the surface of the lenses.

The foregoing remarks on the fitting of spectacles and centring of lenses does not, of course, take into account any prismatic correction that may have been incorporated in the prescription. The centring and fitting of complete spectacles should always be checked in conjunction with the original prescription. Extra precautions will also have to be taken when bifocals are fitted, ensuring that the segment heights are equal and correctly centred on both distance and reading portions.

Normally the glasses will be fitted as close to the face as possible, usually about 12.5 mm from the cornea; this, however, may be varied considerably according to the facial features, length of eyelashes, etc. Where the lenses are of high power, this distance may be critical, and the author advocates that the distance at which the prescription has been obtained should always be specified, so that the prescription can be modified if the glasses have to be fitted in a different plane. It is well to remember that moving a 10 dioptre lens 10 mm makes an effective power-difference at the eye of approximately 1 dioptre. Provision has been made in the latest hospital prescription forms for the back-vertex distance (lens to eye distance) to be specified.

Lenses for ophthalmic use

Compound lenses are never used for ophthalmic

use except in specialized appliances, such as low visual aids, which will be dealt with in chapter 18. Lenses for general use in the correction of ametropia are of a simple positive or negative spherical type, and in many instances are also combined with a cylindrical element for the correction of astigmatism. These lenses are made in many different forms, and are termed: flat, meniscus, toric, bifocal, and lenticular. Other types of lenses may from time to time be produced for specialized purposes.

The above lenses may be described as follows:

Flat lens. A lens either with exactly equal front and back curvatures or, as in most ophthalmic lenses, a spherical curve on one surface and a cylindrical curve on the other.

Meniscus lens. A spherical lens, formed with a concave curve on the inside of the lens and a convex curve on the outside.

Toric lens. The forming or bending of meniscus lenses greatly decreases the aberration of the rays passing through the periphery of the lens, so producing a clearer image. Either of the curves in a meniscus lens may have a cylindrical element incorporated to produce a 'toroidal' surface, hence the term 'toric lens'. This is a technical term from the optical manufacturing industry. A more general term is 'best-form lenses', which may be used to cover any type of lens where the curvature has been calculated to produce the least distortion for a given power. Generally speaking most toric lenses have a curvature of plus or minus 6 dioptres, being the standard base curve, which can be modified according to the required lens power.

Multifocal lenses

As the name implies, bifocal lenses have two different powers, the one in the upper portion being used for near work. Trifocal lenses, which give an intermediate focus, are also in use. For specialized uses the near portion may be placed at the top of the lens; the relative sizes of the distance and near portions may be also varied considerably.

Varifocal lenses have recently been introduced into the ophthalmic range of lenses. As the name implies, these lenses have a variable focus which starts just below the central distance portion and steadily increases to the lower portion of the lens, giving an increased power as the eye scans downwards. These lenses are proving very successful, though initially one has to get used to a certain amount of lateral distortion.

Lenticular lenses are used for high-power prescriptions only. The centre portion of the lens is made to the necessary power; then the outer portion of the lens is ground with a different tool, which considerably reduces the weight; finally the edged thickness is either diminished for negative lenses, or the centre thickness diminished for positive lenses.

The foregoing ophthalmic lenses are in fairly common use, and many other types of lens are made for special purposes, e.g. frosted lenses for occlusion purposes and chevasse lenses, which have irregular surface, to break up the image but at the same time give a cosmetically better appearance over frosted lenses.

Tinted lenses

All ophthalmic lenses may be made up in tinted form, the type of tint varying considerably according to the use to which the spectacles are being put. The lenses may be either surface-coated or tinted in the mass of the lens material. Modern techniques of surface coating are being increasingly used, since if high-power lenses are coloured in the mass the tint will vary considerably according to the actual thickness of the glass, which of course varies according to the power. For some years now tinted lenses which vary in density according to the incident light have been available. Although the early versions took some while to react, modern variable-tint lenses react much more rapidly and are being used more and more in normal spectacle prescriptions.

Size lenses

A brief mention must also be made of special lenses called 'size' lenses for aniseikonia, which within limits can be made to magnify the image and equalize the image disparities between the two eyes. Special reference should be made to published works on this subject.

Prismatic lenses

Any prescription for ophthalmic lenses may require

a prismatic correction to be incorporated. The simplest way of producing a prismatic correction is to move the optical centre of the lens away from the correctly centred position in the spectacles. In this way the relative prismatic effect of the lenses between the two eyes may be adjusted within limits according to the size of the lens available, the amount of prismatic correction required, and the power of the lens.

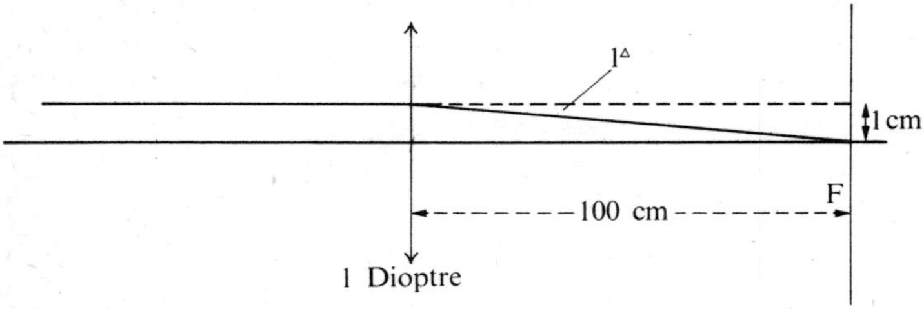

FIG. 17.1.

It will be seen from Fig. 17.1 that a ray of light situated 1 cm from the optical centre of a 1-dioptre lens will be refracted, or bent, one prism dioptre. The following formula may be applied to calculate the prismatic effect of a lens according to the power and decentration: $P = C \times F$, where P is the prismatic effect, C is the decentration in centimetres, and F the power of the lens in dioptres. The amount of decentration will, of course, depend on the diameter of the piece of glass from which the finished lens is to be cut. Where the prismatic effect cannot be obtained by decentration, the lens is supplied with the two surfaces angled at approximately the required prismatic correction, final adjustment being made when the lens is cut from the finished blank. The finished lens will then be equivalent to part of a very large lens with a big decentration.

When a prismatic effect is incorporated in a prescription, the direction of the base has to be specified, and may be expressed in the same term as those used for the cylinder axis. The present trend is for modern spectacles to have very large lenses which are not geometrically centred on the patient's interpupillary distance. More lenses will therefore have to be specially worked where prismatic corrections are incorporated.

Standard notation

When a cylindrical lens is used in the correction of ametropia, it is necessary for the prescription to specify the direction along which the axis of the cylinder must be placed. The power meridian could be specified, but in practice the cylinder axis or meridian of no power is always used. In the past many methods of specifying the direction of the axis have been used, but the Standard Notation system is now used in most countries, and is universally accepted. The position of the axis of a cylinder is always specified with the lens in position as it would be on the patient's face. The 180° line is horizontal, and the intermediate positions are specified, as with a protractor, from 0° to 180° measured in an anticlockwise direction for either eye.

The fitting of spectacles

Further Reading

Bennett, A. G. (1968). *Emsley and Swaine's Ophthalmic lenses.* Butterworths, London.

Bennett, A. G. (1969). *Ophthalmic prescription work.* Butterworths, London.

18 OPTICAL AIDS FOR USE IN SUB-NORMAL VISION

If, after refraction, the visual acuity of a patient cannot be corrected satisfactorily, i.e. where the patient cannot see well enough to follow his occupation or finds the reading of small print impossible, some form of optical aid must be tried and if necessary prescribed. There is a wide variety of optical aids available, ranging from simple hand-held magnifying glasses of various strengths to special spectacles with aspherical or compound magnifiers of widely varying powers. It would be quite impossible to describe the complete range of these special appliances here; and therefore I will briefly outline only the basic principles of the optical system and their uses and difficulties. A number of textbooks have been written dealing fully with this subject.

The simplest form of optical aid, namely the magnifiers, have optical systems exactly the same as those dealt with in Chapter 7 (on lenses) and Chapter 15 (optical principles of ophthalmic instruments). A magnifying glass may be found invaluable for elderly patients, who usually prefer a simple appliance to a complex, but more effective one. Magnification of approximately up to four times may readily be obtained, together with a fairly wide field of vision, without unduly increasing the size of the lens. The main drawback of this type of appliance is that it is difficult to hold steady and therefore not always suitable for the 'shaky' patient; it also entails the use of one hand to hold it, thus restricting the number of occupations for which it can be used. This last difficulty may be overcome to some extent by using a suitable stand.

If special magnifying spectacles are made, 'best-form' lenses are very often employed in order to obtain a wider field of vision without distortion. At least one surface of these best-form lenses is usually made aspherical. The main drawback of this type of appliance is the very short working distance they allow. A lens of 20 dioptre would mean a working distance of only 5 cm. One advantage, however, is that a reasonable field of vision can be obtained with this appliance, and since it is in the form of spectacles both hands are free.

Where a high magnification with greater working distance is required–and in fact in most cases this is desirable–it can only be obtained with some form of compound optical system, which can be made to produce an overall magnification and at the same time be adjusted for viewing a near object. The Galilean telescope is normally used in one form or another in this type of appliance, and consists of a positive spherical lens object-glass and a negative eyepiece (see Fig. 18.1). This simple instrument produces a magnification equal to the ratio of the two focal lengths of the object-glass and eyepiece, the two lenses being separated by a distance equal to the difference between these two focal lengths. It will readily be seen that a very small and compact optical system producing a magnification of approximately two times can fairly

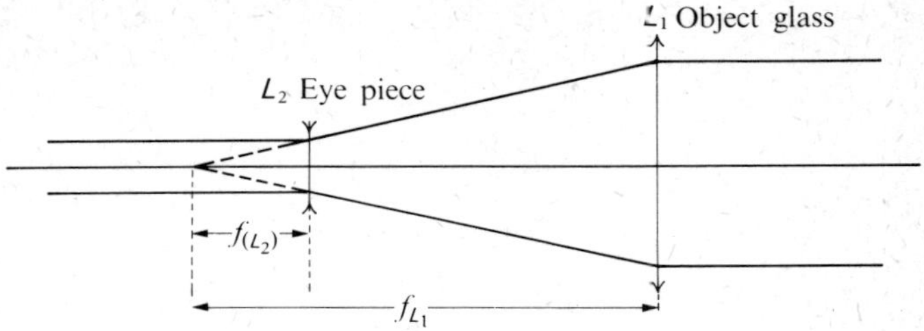

$$\text{Magnification} = \frac{f_{L_1}}{f_{L_2}}$$

FIG. 18.1. Galilean telescope.

easily be produced. Magnifications higher than two times become difficult to manufacture if a good optical system is important as well as a reasonable field and an absence of distortion. This optical aid in a Galilean form would normally be focused at infinity, but by increasing the power of the object-glass a near object would be brought into focus, while still producing an overall magnification approximately two times greater than that produced by a simple magnifier focused at the same working distance. The ratio of the size and power of the lens vary considerably from one manufacturer to another. This type of appliance may be obtained in powers ranging from approximately two to eight times, but are normally used binocularly only in powers of magnification of up to approximately four times. Where higher magnifications are required for binocular use, special telescopic magnifiers can be used in which each element is mounted in a special frame angled to coincide with the optical axis. When high-power magnifiers are used the difficulty of adequate lighting arises, and some appliances have a built-in illuminating device incorporated.

No mention has been made so far of optical aids for distance vision, and at best these can be made only to a magnification of roughly two times. The field loss on this type of appliance is so great, however, that in the author's opinion they can be dangerous for general use, and in fact they are rarely supplied; the chairbound, however, may find them a considerable help at home for television-viewing, etc. Monoculars and binoculars with higher magnifications may also prove of value for occasional use.

The prescribing of these special appliances requires considerable experience and patience. Many factors, such as age, occupation, etc., must be taken into account in an endeavour to find the best optical aid possible. Mention must also be made of the psychological effect on the patient of even the smallest help that can be given.

Further Reading

Bier, N. (1970). *Correction of subnormal vision* (2nd edn). Butterworths, London.

19 OPTICAL PRINCIPLES OF OPHTHALMIC INSTRUMENTS

The optical principles of the ophthalmoscope and retinoscope were dealt with under the appropriate chapters, but there are a number of other optical instruments used in the examination of the eyes, and a brief mention of the principles of simple magnifiers and slit lamps will probably be of value.

When viewing any part of the eye some form of magnification is invariably required. In most specialist instruments, i.e. ophthalmoscopes and slit lamps, this is obtained with some system of magnification, such as a microscope, and in many instances it is of a compound optical system which may be varied for magnification. The hand-magnifier or loupe is a ready means of obtaining a fairly high magnification for most purposes, being simple to use and small to carry about. These hand-magnifiers usually have fully corrected lenses from the optical point of view, to obtain as wide and clear a view as possible.

The amount of magnification produced will, of course, vary according to the distance at which the magnifier is held from the object being viewed; a universal method for standardizing the magnification is therefore usually adopted. The magnification when using a loupe is the ratio of the maximum size of the object without the use of the glass to the size when using the glass. As the maximum size without the glass will vary according to the distance at which it is placed from the eye, a standard distance of 250 mm (the distance of distinct vision) is normally adopted. When using a magnifying glass for maximum magnification the object is normally placed at or near the focal point of the lens. As shown in Fig. 19.1 the overall magnification will then be equal to

$$\frac{\text{Distance of distinct vision (250 mm)}}{\text{Focal length of magnifier in mm}}.$$

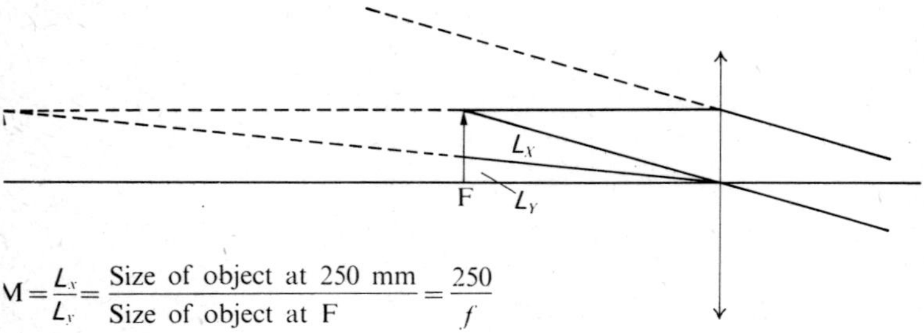

$$M = \frac{L_x}{L_y} = \frac{\text{Size of object at 250 mm}}{\text{Size of object at F}} = \frac{250}{f}$$

FIG. 19.1. Simple magnifying glass.

Magnifiers and loupes are sometimes made in a binocular form, being attached to a head band, or made in the form of spectacles. The magnification would normally apply in the same way, but a lower power has to be used as it would be impossible to converge on to an object if a high-power binocular magnifier was used. In most low-power binocular magnifiers some form of prism is incorporated to assist in the convergence. If a higher power is required a prismatic binocular loupe may be used wherein the light from the object is reflected by the means of prisms before entering the eye. The emergent beams are adjusted for the examiner's inter-pupillary distance (see Fig. 19.2).

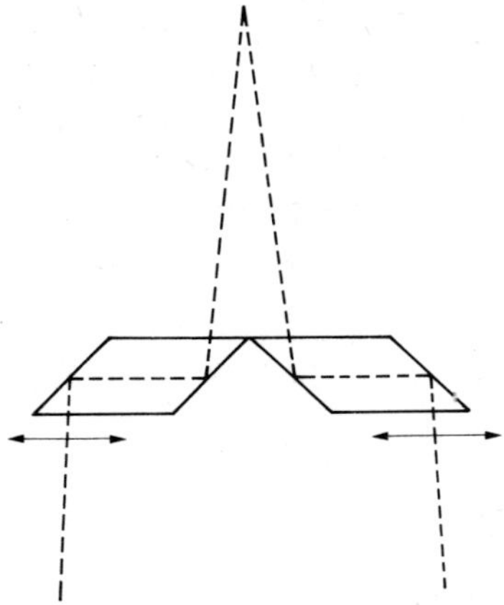

FIG. 19.2. Prismatic binocular loupe.

Where a higher magnification is required a different form of instrument with a compound or two-stage magnification is necessary. The slit lamp incorporates the principle in the microscope. The optical system of all microscopes is basically the same but is modified according to the needs for which the instrument is designed. From Fig. 19.3 it will be seen that an enlarged image of the object is formed at B by the objective X in the focal plane of the eyepiece. This object is then viewed by the eyepiece (a simple magnifying glass), producing an overall enlarged image depending on the various optical elements and their separation. From this diagram it is seen that the

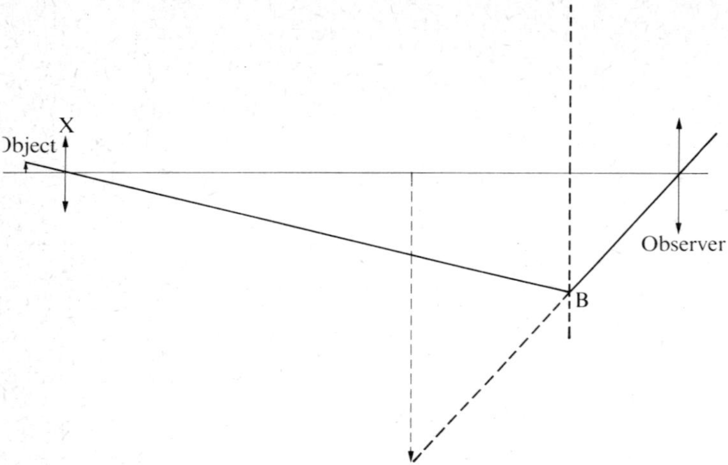

FIG. 19.3. Optical system of microscope.

image would be inverted which would, of course, be unacceptable in most ophthalmic instruments so some method of erecting the image must be employed. This is normally done by means of prisms being inserted between the objective and the eyepiece. These prisms not only erect the image but considerably reduce the distance between the objective and the eyepiece, making a more compact intrument. Most modern slit-lamps have some means of varying the overall magnification, either by several intermediate stages or by incorporating zoom lenses.

The illumination system of the slit lamp is of a similar principle to that of the normal projector, but having the object and the image distances reversed, the image distance being only a few centimetres. Fig. 19.4 shows the outline of the optical path. The fila-

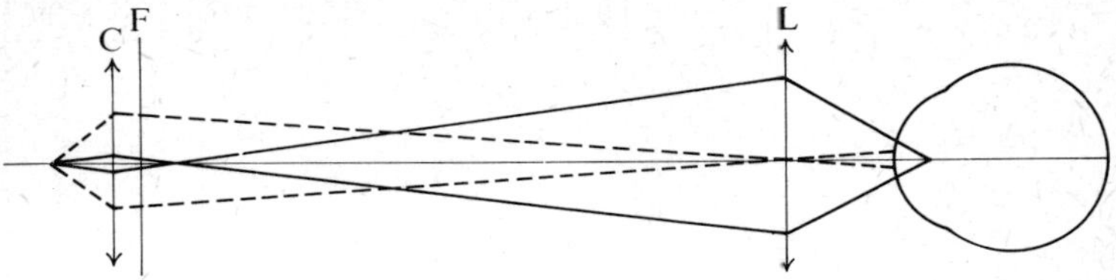

FIG. 19.4. Illumination system of slit lamp.

ment of the lamp is imaged by condenser C on to the projector lens L. The condenser ensures the slit F is brightly and evenly illuminated and the whole system is rather like a slide projector. The slit is then imaged by the projector lens in the plane of the eye under observation. Most slit lamps today employ coincidence setting of the illuminating beam and visual focus. The visual focus and slit image may then be angled relative to each other about a common point situated in a position approximately below the centre of rotation of the eye.

As with other optical instruments, the relative positions of the lenses and foci are varied from manufacturer to manufacturer, but the basic principle remains the same.

Further Reading

Fincham, W. H. A. and Freeman, M. H. (1974). *Optics* (8th edn). Butterworths, London.

20 CONTACT LENSES

Contact lenses have been used throughout the world for many years, and a book of this sort would not be complete without some mention of their optical principles and also a brief mention of their possibility for their use therapeutically.

As the name implies, contact lenses are lenses close to or in contact with the eye, and from the lay point of view their value is probably best known as a means of assisting vision without the use of normal conventional spectacles. Great advances have been made in the past ten or twenty years in the types of lenses available and the materials and technqiues from which they are made. It would be necessary to refer to complete works on this particular subject to understand the whole of the techniques of fitting and optical theory, only a brief resumé, therefore, will be attempted to try and give the reader an outline of the types of lenses which can be obtained and their optical principles.

We have seen in earlier chapters that the refraction of the eye is dependent not only on the length of the globe but also on the curvature of the front surface of the eye, and these combine to form an optical system whereby, if the rays of light do not come to an accurate focus on the retina the patient will have blurred vision. Conventional spectacles will merely alter the position of the rays of light inside the eye until the correct requirements are satisfied, namely an accurate focus on the retina. With a contact lens a simple analogy is merely to change the anterior curvature of the cornea by means of a lens placed on, or in close proximity to, the cornea. In the case of hard contact lenses the curvature of the anterior surface of the optical system of the eye is thus altered in conjunction with a lenticular tear-film between the contact lens and the cornea. Soft contact lenses, which are in actual close contact with the cornea, will be mentioned later.

The anterior and posterior curves of contact lenses are generally not equal. The anterior surface is modified to correct the refractive error whilst maintaining the posterior curve for fitting the cornea correctly.

Broadly speaking there are two types of contact lenses, corneal and scleral. The scleral lens is a large lens having two zones, the scleral and corneal zones. The scleral zone is an annular ring which conforms roughly to the curvature of the sclera in the centre of which is the corneal portion, on which is worked any optical correction necessary. These lenses are rather bulky and cannot always be tolerated for long periods.

Recent techniques have developed the corneal lens, which has no scleral portion and 'rides' on top of the cornea with a very faint tear-film beneath. It is very much smaller than the scleral lens, usually smaller than the cornea itself. Earlier contact lenses, particularly the scleral lenses, were made from glass, but practically all contact lenses are now produced in some form of plastic material which can be moulded very accurately, and worked and polished where necessary to get the correct power.

In recent years an entirely new form of contact lens has been produced, namely the soft lens, which is produced from a soft material which when placed in contact with the cornea moulds itself to the surface curvature of the cornea. These lenses are now being widely used although the actual technique and the efficiency of the lens is still subject to controversy.

The cornea, being a living tissue, has to breathe. Oxygen intake has to be correct and one of the difficulties with contact lenses is the probable upsetting of this balance of the ventilation of the cornea. The early scleral contact lenses (which were usually made of glass) had to be fenestrated if they were to be tolerated for long (a small hole being pierced through the lens). This allowed a ready exchange of the tear film, with the consequent carrying through of new oxygen through this small hole, and the lenses could be worn for much longer periods. With the advent of corneal lenses, this question rarely arises because there is, or may be, a constant flow of tears between the surface of the contact lens and the cornea. The correct fitting of any contact lens is dependent on the accuracy and skills of the contact lens practitioner, although with the soft lenses a different technique is involved. The lenses have some permeability to fluids and the oxygen breathing effect on the cornea can therefore be assisted. If this type of lens is permeable to fluids, they may also be permeable to any bacteria which may be present, making sterilization more difficult. The development of various types of material for soft contact lenses is still under experimentation, so as to produce a lens which can safely be sterilized in a simple way.

Some of the more recent techniques consist of using materials enabling the lens to be worn permanently and only removed at rather long intervals. This latter type of lens is being produced under various trade names, but is generally referred to as 'extended wear' lenses.

As with all optical aids astigmatism is a problem in contact lens prescriptions. If the astigmatism is largely corneal, much of the astigmatism will be neutralized by the tear film when fitting the contact lens, with the exception of soft lenses which mould and include the astigmatism in the final surface when in contact with the eye. In some instances with hard lenses, astigmatism can be worked into the lens and so correct the optical astigmatism in the same way as with normal spectacles. Where there is high corneal astigmatism, fitting becomes more difficult and it is sometimes essential that the scleral contact lenses are fitted. Most scleral contact lenses these days are fitted to the eye by a moulding technique which fits the scleral portion of the lens very closely to the contour of the eye, and will therefore be much less likely to revolve round the optical axis, thus keeping the cylindrical axis in the correct place. With soft lenses a compromise has to be made to produce the best possible vision, leaving some residual astigmatism which may possibly be slightly lower than the normal vision obtainable with glasses. Recent advances, however, in the manufacture of soft lenses have resulted in some toric surfaces being produced.

Contact lenses may be aesthetically required, or optically required, or a combination of both. This is something the patient and his practitioner must decide. There are, however, some instances in which contact lenses are a great advantage, namely in the high myopic or high hypermetropic patient where the inevitable distortion through the edges of the lenses may be severe with conventional spectacles. In these instances contact lenses, where tolerated, can be extremely advantageous. Another advantage in the use of contact lenses is the case of the unilateral aphakic patient, where the phakic eye is still giving normal satisfactory vision. The patient would be unable to fuse the two images with conventional spectacles due to the increase in the image size in the aphakic eye, but with a contact lens the image-size difference is considerably reduced and the patient may well be able to tolerate the two eyes in a binocular fashion.

Where there has been damage to the anterior surface of the eye due to ulceration or other medical reasons, leaving an irregular cornea, a distorted image may result which cannot be corrected by normal lenses. Contact lenses will virtually provide a new anterior surface to the cornea, and a considerable increase in vision might then be obtained.

Apart from instances of this sort, there are certain therapeutic uses of contact lenses which will be mentioned for interest:

(1) Contact lenses can be used for the retention of certain drugs being used for treatment of the eye.

(2) Specially prepared contact lenses for ptosis props have proved very successful.

(3) The provision of contact lenses in keratoconus is essential as there is sound evidence that they delay the progression of this disease.

(4) Soft lenses are also proving invaluable as 'bandage' lenses often saving resort to a tarsorrhaphy.

The use of contact lenses today is without a doubt fully established and the techniques of fitting, the types of material used, and methods of manufacture will continue to advance in the years to come.

Further Reading

Stone, J. and Phillips, A. J. (1972). *Contact lenses*. Barrie and Jenkins: Butterworths, London.

BIBLIOGRAPHY

British Journal of Ophthalmology (1974). Strabismus Symposium. *Brit. J. Opthalmol.* **58**.

Burian, H. and Von Noorden, G. K. (1974). *Binocular vision and ocular motility*. Mosby, St. Louis.

Cashell, G. T. and Durran, M. (1975). *Handbook of orthoptic principles* (3rd edn). Churchill Livingstone, Edinburgh.

Duke Elder, S. (ed.). *System of ophthalmology*, Vol. 6: *Ocular motility and strabismus*. Kimpton, London.

Orthoptics (Proceedings of the Second International Orthoptic Congress, 1971). Excerpta Medica, Amsterdam.

Orthoptics (Transactions of the Third International Orthoptic Congress, 1975). Stratton Intercontinental Medical Book Corporation, New York.

Transactions of First International Congress of Orthoptists (1968). Henry Kimpton, London.

INDEX